Beyond the Green Revolution: New Approaches for Third World Agriculture

Edward C. Wolf

Worldwatch Paper 73
October 1986

Sections of this paper may be reproduced in magazines and newspapers with acknowledgment to Worldwatch Institute. The views expressed are those of the author and do not necessarily represent those of Worldwatch Institute and its directors, officers, or staff.

Library of Congress Catalog Card Number 86-51251
ISBN 0-916468-74-7

Printed on recycled paper

Table of Contents

Introduction

5

After twenty years, the green revolution stands as a touchstone in international agricultural development. At a time when famine seemed imminent, new varieties of wheat and rice introduced to Asia and Latin America along with fertilizers, pesticides, and mechanized farm equipment dramatically increased harvests. This agricultural strategy, which transformed the lives and prospects of hundreds of millions of people, is considered the most successful achievement in international development since the Marshall Plan and the reconstruction of Europe following World War II. India, whose food prospects once seemed bleak, today holds grain reserves that provide insurance against famine. Indonesia, once the world's largest rice importer, is now self-sufficient and exports rice.

But the agricultural progress that made the green revolution possible has not been distributed evenly. New seeds, fertilizers, and pesticides boosted the crop yields of Asian and Latin American farmers who had access to irrigation systems and markets for their crops. The aggregate statistics hide a large group of Third World farmers who did not benefit from the new technologies: subsistence farmers raising food for their families on marginal, rainfed land. Because their agriculture remains unproductive and vulnerable to crop failure, drought, and natural catastrophe, these rural people remain among the poorest in their societies. Failing to address their needs has slowed economic progress in dozens of countries. The recurrent famines in Africa, and persistent pockets of starvation on that continent, demonstrate the unacceptable human costs of this neglect.

I sincerely thank Alan Durning for research assistance, Susan Norris for production assistance, and Maureen Hinkle, Bruce Horwith, William Liebhardt, Robert Rodale, Mark Silberman, John Stuyt, Ed Sultzberger, Ken Tull, and Garth Youngberg for their helpful comments on an early draft of this manuscript.

High-yielding varieties of wheat and rice have been introduced to less than a third of the 423 million hectares planted to cereal grains in the Third World. The rates of adoption vary widely by region: 36 percent of the grain area in Asia and the Middle East, 22 percent of Latin America's grain area, and only 1 percent of Africa's grain fields grow improved varieties of wheat and rice. Other crops including barley, sorghum, potatoes, and especially maize have also been improved by research and breeding, and new varieties distributed to farmers. The local contributions of these advances have been substantial. For example, Zimbabwe's maize surpluses in recent years stem largely from plantings of improved hybrids by commercial and communal farmers alike. But none of these crops have had an effect on total food production, average productivity, and rural incomes as widespread or significant as the green revolution wheats and rices.[1]

Not all wheat and rice farmers can afford the new seeds and the inputs they require. Others raise crops for which systematic research is just beginning. Overall, nearly 100 million people in Latin America, 280 million in Africa, and over 990 million in Asia raise food under difficult conditions at yields little changed since mid-century. But grain yields in more agriculturally advanced regions are already near their biological ceilings; thus this group of nearly 1.4 billion people holds the key to future increases in world food production.[2]

The case for increasing yields remains as compelling today as it was a generation ago. Over the next 13 years, world population will expand from today's 5 billion to over 6 billion. Few analysts expect a significant expansion of cultivated land by then. Just to maintain current consumption levels will require a 26 percent increase in the world's average grain yields. And by 2020, feeding the projected population of 7.8 billion will require grain yields 56 percent higher than 1985 levels. Unlike past yield increases achieved under favorable cropping conditions, future improvements in average yields must come from raising the productivity of traditional farmers who cultivate low-yield crops under marginal conditions—perhaps the most demanding challenge that national governments and the international development community have faced.[3]

"High-yielding varieties of wheat and rice have been introduced to less than a third of the 423 million hectares planted to cereal grains in the Third World."

Small farmers cultivate their crops under extraordinarily diverse ecological conditions, ranging from the rain-soaked volcanic archipelagos of Southeast Asia to the arid savannas south of the Sahara and Latin America's high altiplano. Farming methods, and staple foods, vary enormously as well. In Southeast Asia, for example, where one-third of the farms are less than half a hectare in size, most farmers depend exclusively on manual labor and draft power supplied by water buffalos to cultivate their rice. On Africa's small farms, cultivated more with hoes than plows, families grow root and tuber crops including cassava and yams as the primary staples. Despite such variety, subsistence farms around the world share common features: Farmers often mix different crops in the same field to reduce the risk if a particular crop fails, they grow a variety of staple crops and vegetables to meet family food needs, and they rarely purchase artificial fertilizers or pesticides.[4]

Green revolution approaches will only be part of the answer for the 230 million rural households in Africa, Asia, and Latin America that use farming methods almost identical to those of their ancestors. One reason is energy. Past advances have come from increasing the energy intensity of farming: fuel to run machinery, fossil-fuel-based artificial fertilizers, and diesel fuel or electricity to run irrigation pumps. Few of the rural poor can afford these costly materials and services. Even if they had the income to purchase such inputs, many farmers are not served by roads or markets that could reliably supply them.

In addition, subsistence farmers grow crops that have received comparatively little research attention. There is as yet no research base for achieving high yields in many staple crops. Third World farmers cultivate on poor soils under harsh climates that require finely tuned agricultural practices. As rural populations grow, these farmers will need agricultural advances that are labor-intensive, rather than capital- and energy-intensive. Such conditions demand different research approaches from those that raised yields in the past.

A new strategy of efficiency and regeneration could help meet the needs of subsistence farmers, and begin to address the environmental and economic problems linked to more intensive cropping practices as well. Such a strategy would stress the efficient use of fertilizers, chemicals, water, and mechanized equipment. As a supplement to efficiency, farmers would blend biological technologies and traditional farm practices to increase the contribution that the land's natural fertility makes to food production.

Two sets of technical opportunities, already stirring the agricultural research community, promise rapid progress toward better resource management and regenerative approaches. One is the reappraisal of traditional farming practices, once judged backward and unproductive. Shifting cultivation, multiple cropping, and traditional soil management methods, though often practiced under pressures that make them counterproductive today, are governed by ecological principles that can serve as models for sustainable farming.

Biotechnology can also offer new solutions to Third World farming problems. The ability to modify the genetic makeup of plants and animals poses environmental risks that must be carefully evaluated. But biotechnology's benefits lie in the potential to allow breeders to develop new crop varieties more quickly than conventional breeding. Crops may be tailored to use water and nutrients more efficiently, and to perform well in the mixed cropping practices that many poor farmers employ.

Joining biotechnologies with the ecological insights of traditional farming promises innovative solutions to agriculture's economic and environmental problems. Government policies do not fully recognize that promise yet. But opportunities have never been greater for moving agriculture toward sustainability and reaching the quarter of the world's people—and quarter of the world's cropland—left out of the green revolution.

Productivity Reconsidered

The pursuit of productivity has been central to agriculture since farmers first selected the wild grasses ancestral to today's crops. In recent years, harvests have outpaced population growth, not only because more land has been brought under the plow, but because different plant varieties, more irrigation, fertilizer, and improved tools and equipment allow farmers to produce more from each hectare of land and each hour worked. World grain production increased from 620 million tons in 1950 to 1,660 million tons in 1985, and the average yield per harvested hectare climbed from 1.1 tons to 2.6 tons. These rapid increases have no precedent.[5]

The postwar increase in yields rested on a simple formula. Researchers and extension agents encouraged farmers to use more fertilizers, pesticides, and irrigation in combination with newly bred crop varieties. According to the conventional approach, substituting these capital- and energy-intensive inputs for the traditional resources of land and labor would allow farmers to expand harvests each year.

Some countries succeeded handsomely; hundreds of millions of people are better fed and better off than they would have been without these gains. But because of their exclusive focus on improving yields, policymakers and researchers emphasized regions where the economic return on investment in fertilizers would be highest, rather than seeking to distribute inputs more widely. This approach naturally overlooks farmers on marginal land, for whom raising yields may not be as important as increasing resistance to drought and other natural catastrophes.

Despite its drawbacks, enthusiasm for the conventional productivity formula is understandable. The increase in world food production in the last decade has been associated with a comparable increase in the use of artificial fertilizers. Regions that have used the most additional fertilizer have reaped the largest benefits. (See Table 1.) Asia and North America, the areas that harvested nearly four-fifths of additional world grain production, accounted for 56 percent of the increase in fertilizer use. And while North America's average harvested area also expanded over this period, virtually all the growth in Asian harvests came from fertilizer. Yet Eastern Europe and the Soviet Union demonstrate that additional fertilizer alone does not necessarily mean proportionately larger harvests. Relying on central planning rather than farmers to allocate fertilizer supplies accounts for much of the inefficiency.

Table 1: Increase in Average Grain Production and Fertilizer Use by Region between 1970-74 and 1980-84

Region	Grain Production: Total Increase	Grain Production: Share of World Increase	Fertilizer Use: Total Increase	Fertilizer Use: Share of World Increase
	(MMT)[1]	(percent)	(MMT)	(percent)
Asia	200.3	55	19.2	45
North America	86.0	23	4.6	11
Western Europe	31.3	9	2.8	7
Latin America	23.9	7	3.2	8
Eastern Europe and Soviet Union	8.8	2	10.7	25
Oceania	8.4	2	0.6	1
Africa	8.2	2	1.5	3
World	366.9	100	42.6	100

[1]Million metric tons.

Sources: U.S. Department of Agriculture, Economic Research Service, "World Indices of Agricultural and Food Production 1950-85," unpublished printout, Washington, D.C., April 1986; U.N. Food and Agriculture Organization, *Fertilizer Yearbook* (Rome: 1982 and 1984).

Average grain yields in the world's most populous countries reflect in part variations in rainfall and soil fertility, but they also illustrate the productivity gap that must be closed in the effort to raise the world's average yield above 2.6 tons per hectare. (See Table 2.) The 11 countries shown in Table 2 are home to nearly two-thirds of the world's

Table 2: Land Productivity in World's 11 Most Populous Countries, 1985

Country	Average Grain Yield	Population
	(tons per hectare)	(million)
Japan	5.8	122
United States	4.8	241
China	3.9	1,050
Indonesia	3.7	168
Bangladesh	2.2	104
Mexico	2.1	82
Brazil	1.8	143
India	1.6	785
Pakistan	1.6	102
Soviet Union	1.6	280
Nigeria	0.8	105
Total Population		3,182

Sources: Population Reference Bureau, *1985 World Population Data Sheet* (Washington, D.C.: 1985); U.S. Department of Agriculture, "World Indices of Agricultural and Food Production 1950-85," unpublished printout, Washington, D.C., April 1986.

population and represent the entire economic and ecological spectrum. Slightly fewer than a third of the world's people live in four countries where land productivity, measured as tons of grain harvested per hectare of agricultural land, exceeds 3.5 tons, well above the world average. Another third live in the five countries where productivity is less than 2 tons per hectare. While the highest yields occur in affluent industrial nations, China and Indonesia demonstrate that low income need not be associated with low yields.

The first step most countries can take to increase harvests is to correct the inefficient application of chemical fertilizers. Even China's high grain yield conceals a substantial opportunity to expand total harvests by distributing fertilizers more equitably to Chinese farmers. China's remarkable increase in food production from less than 200 million tons in the mid-seventies to over 300 million tons by 1985 was made possible in large part by an equally dramatic increase in fertilizer use. By 1983, Chinese farmers were applying 115 kilograms of artificial fertilizers per hectare planted, about as much as U.S. farmers. But according to Bruce Stone of the International Food Policy Research Institute, most of this was destined for just a third of Chinese cropland, that in the country's most fertile and most market-oriented areas. Adding another sack of fertilizer to these fields now produces much less additional grain than fertilizing neglected areas. Distributing fertilizer to the other two-thirds of Chinese cropland could yield three to 15 times more grain per ton of additional fertilizer than the state and market-oriented farms could produce under the existing distribution system.[6]

In addition, farmers achieve less than optimum production when they apply the nitrogen, phosphorus, and potash in artificial fertilizers in incorrect ratios. Nitrogen is often in short supply in Chinese soils, but if other nutrients are also lacking, just adding nitrogen cannot raise yields. Addressing the inequities in fertilizer distribution and correcting nutrient imbalances can increase food output faster and at lower cost than simply expanding fertilizer supplies, even in countries that now use little fertilizer.[7]

Another reason for using fertilizer more efficiently is the high environmental costs linked to heavy use. Government subsidies in Europe, Japan, and North America encourage farmers to expand production by applying more fertilizer than either sound agronomic practices or world market conditions warrant. One result of this subsidized overfertilization is that as much as one-fourth of the nitrogen fertilizers used in these regions leaches into groundwater. Increasing concentrations of nitrates in drinking water, which pose a health threat to bottle-fed infants, have been reported in Denmark, England, France, Germany, and the Netherlands. Ironically, at the levels of fertilizer applied by European farmers, the losses of nitrogen may

"The first step most countries can take to increase harvests is to correct the inefficient application of chemical fertilizers."

amount to 30 to 45 kilograms per hectare—more fertilizer than is applied to cropland in many Third World countries.[8]

Farmers in Africa, Latin America, and Oceania have used the least additional fertilizer and contributed least to expanded food supplies. In Latin America, the challenge of managing enormous external debts has forced many countries to curtail imports of fertilizers in an effort to conserve foreign exchange for interest payments. In Africa, few farmers can afford conventional fertilizers, and limited water supplies often make them unprofitable. Yet, African and Latin American farmers need to expand food production, which has fallen behind population growth in both regions. Using additional fertilizer more efficiently would help, but these farmers also need less costly alternatives to the conventional methods of raising productivity.

Correcting inefficiencies in the use of purchased resources is not the only way to raise and sustain agricultural productivity. As the circumstances facing Africa and Latin America suggest, helping farmers achieve more stable yields, manage soils and water supplies more effectively, and control spending on costly chemicals could make farming practices in many settings more sustainable. Redefining productivity to encompass these concerns could broaden options for poor farmers in developing countries and suggest new directions for financially strapped farmers in industrial countries.

Farmers have another set of productive resources that publisher Robert Rodale has aptly labeled the "internal resources" of agriculture: the inherent fertility of the soil, rainfall and climate patterns, the dynamics of pest populations and their natural enemies—in other words, the natural resource base. The productive potential of these internal resources is sometimes masked or even diminished by heavy use of artificial fertilizers and other farm chemicals.[9]

"The rapid introduction of external inputs into agricultural production over the past century has unnecessarily diminished the strength, vitality, and usefulness of the internal resources of farmers," Rodale argues.[10] Research on nitrogen fixation by legumes shows how this can happen. Microorganisms in the roots of these crops convert nitrogen from the air into a form plants can use; the excess that remains in the soil can help nourish a subsequent grain crop. Soil scientist David Bezdicek and his colleagues at Washington State University have found that residual nitrogen from artificial fertilizer can reduce the amount of nitrogen fixed by a legume crop such as chickpeas. A heavy dose of fertilizer applied at the start of the growing season suppresses biological activity, while in some cases a small amount of fertilizer can actually stimulate nitrogen fixation. More nitrogen might be supplied by the correct balance of artificial fertilizer and biological nitrogen fixation than by using artifical fertilizers alone.[11]

The regenerative approach seeks to maximize biological contributions to agricultural productivity. It makes the most of natural sources of nitrogen, phosphorus, and potash, as well as the way these nutrients are cycled and conserved in natural ecosystems. Regenerative farming practices include sowing different crops together to use fully the soil's fertility, rotating food grains with nitrogen-fixing legumes, and planting trees and shrubs whose roots draw nutrients from deep soil layers to the surface. Purchased fertilizers and pesticides are used sparingly in these practices. Although regenerative methods require more careful farm management, they are less costly than conventional approaches.[12]

Agricultural research that emphasizes biological approaches to raising productivity can help poor farmers better cope with the risks imposed by erratic rainfall and less fertile soils. Conventional agricultural modernization, based on fossil fuels, is already beyond the means of many Third World farmers. Offered better methods for managing their internal resources, these farmers can reduce their vulnerability to crop failure and famine.

Beyond the Green Revolution

Two decades have passed since new, high-yielding varieties of wheat and rice were introduced to farmers in Mexico, the Middle East, and South Asia. The new varieties, which were more responsive to artificial fertilizers and irrigation than traditional varieties, "spread more widely, more quickly, than any other technological innovation in the history of agriculture in the developing countries."[13]

"Research that emphasizes biological approaches can help poor farmers better cope with erratic rainfall and less fertile soils."

Modern grain varieties were quickly taken up in Bangladesh, India, Pakistan, and throughout Southeast Asia. (See Figures 1 and 2.) In Latin America, the area planted to new wheat and rice varieties increased from 270,000 hectares in 1970 to 9.6 million hectares by 1983. By the mid-eighties, roughly half the wheat area and nearly 58 percent of the rice area of all developing countries had been sown to high-yielding varieties. In major wheat- and rice-growing regions, the percentages are far higher: 82 percent of Latin America's wheat area and 95 percent of China's rice area were sown to high-yielding varieties in 1983.[14]

The amount of rice and wheat grown in developing countries increased 75 percent between 1965 and 1980, while the area planted to those crops expanded by only 20 percent. The ability to harvest two crops a year with the new seeds contributed to these increases. In 1980, the additional wheat and rice produced by green revolution technologies were worth an estimated $56 billion, of which $10 billion was due to the improved genetic potential of the new varieties. This expansion of the food supply is crucial to many countries with rapidly growing populations. As Michael Lipton of the University of Sussex wrote in 1985, "If the farmers of the Third World today used the same cereal varieties as in 1963-64, and everything else were unchanged, then tens of millions of people would this year die of hunger."[15]

Africa benefited least from the green revolution. Few of Africa's 50 million rural families grow wheat or rice, and only in the last decade have researchers turned their attention to millet, sorghum, cassava, yams, and cowpeas that are the subsistence staples of most rural Africans. Only 6 percent of sub-Saharan Africa's wheat and rice area is planted to modern varieties. Improved maize varieties and hybrids have boosted harvests in countries including Kenya, Zimbabwe, and South Africa, but on the whole, scientific plant breeding has not decisively changed the continent's food prospects. And Africa's conditions are not unique; many farmers in Latin America and Asia are still prevented from planting improved varieties by poor soils, erratic water supplies, and poverty.[16]

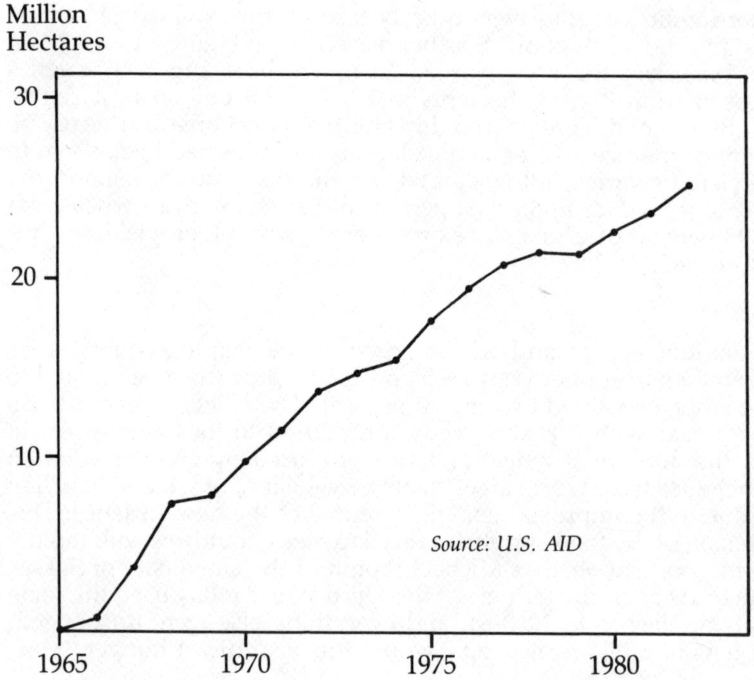

Figure 1: Area Planted to High-Yielding Varieties of Wheat in South Asia, 1965-82

High-yielding varieties of wheat and rice are still spreading, however. Though the early green revolution seeds were planted almost exclusively by farmers with well-irrigated land who could afford to purchase the necessary supplements of fertilizers and pesticides, modern varieties are now grown by farmers under less-favored circumstances. More than half of the high-yielding wheat in Bangladesh is watered only by rain, as is about 85 percent of the high-yielding rice in the Philippines. Varieties bred and released today perform better than traditional varieties even without costly inputs.[17]

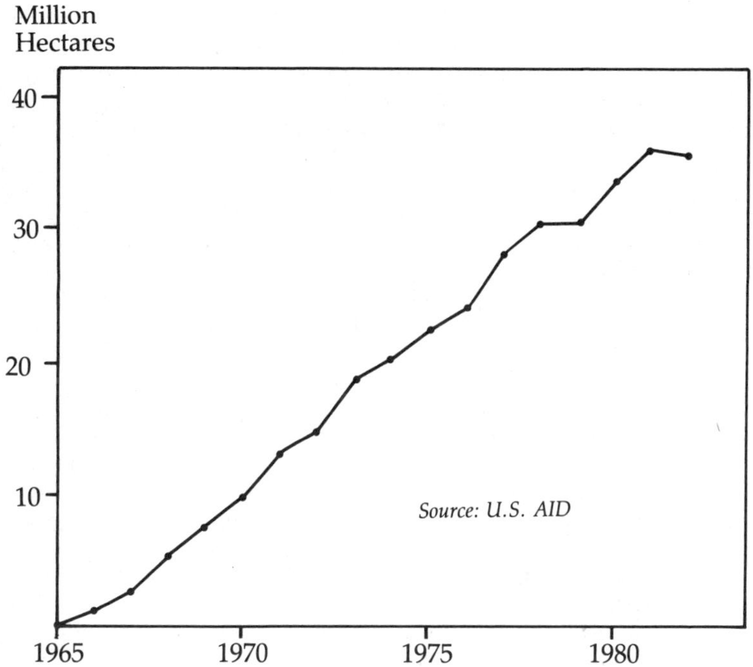

Figure 2: Area Planted to High-Yielding Varieties of Rice in Southeast Asia, 1965-82

The green revolution's early benefits were by no means equally shared. Relatively prosperous farmers who controlled more land, and so had the financial means to purchase fertilizers, pesticides, and equipment, gained most by adopting high-yielding wheat and rice. Small farmers in areas favored by abundant water, who tended to adopt new varieties and technologies later, also profited, but not as much. Grain prices dropped because of bigger harvests on the larger landholdings. Consumers enjoyed the lower food prices—or at least the brake on price increases—that expanded harvests made possible.

Governments also gained. India, for example, used the expanded production of the late sixties to offset its dependence on costly grain imports rather than to significantly increase food consumption among its poor. A drain on the country's treasury was plugged but there was little progress in raising per capita food consumption.[18]

Others, however, lost from the new technologies. The biggest losers were farmers in areas where the new varieties performed poorly, and those growing crops primarily for subsistence. These farmers earned no new income from bigger harvests, and may have become poorer as prices for their occasional marketable surpluses declined. On some 300 million hectares in the Third World, supporting over a billion people, productivity has not measurably improved.[19]

That the record of the green revolution is mixed should come as no surprise. The scientists who developed the new varieties of wheat and rice never expected their work to provide an open-ended solution to the world's food problems. Many believed that the new technologies offered a means to buy time until population growth rates could be slowed. Harvests could not increase indefinitely; birth rates would have to fall. Twenty years later, countries like China that both promoted new seeds *and* instituted economic reforms and national family-planning programs to lower birth rates have done the most to improve the welfare of their people.

The unique research network launched by the Rockefeller Foundation in Mexico in 1943 may be a more significant contribution of the green revolution than the expanded harvests achieved so far. Supported by the Rockefeller and Ford Foundations, plant breeders developed new crop varieties appropriate for conditions in Mexico, Pakistan, India, and Turkey. Success in these countries led to the creation of the Philippines-based International Rice Research Institute (IRRI) in 1962, the International Center for the Improvement of Maize and Wheat near Mexico City in 1965, and ultimately to the creation of a system of 13 international agricultural research centers funded through the Washington-based Consultative Group on International Agricultural Research (CGIAR). The research agenda of the centers today covers 21 food crops, conservation of the genetic resources used for plant

"Scientists never expected their work to provide an open-ended solution to the world's food problems."

breeding, animal husbandry and livestock diseases, and policy issues related to agricultural research.[20]

A high priority of the CGIAR centers is the need to defend the gains already achieved. Farmers who plant high-yielding varieties of wheat and rice need continuous research to sustain their yields. This "maintenance research" emphasizes breeding crop varieties to increase their natural ability to resist pests and disease. Maintaining stable yields at high levels can be a more complex task than raising yields in the first place. New plant varieties must be on hand to replace old ones that succumb to pests and disease. This requires a vast breeding program and an extensive system of gene banks.[21]

National and international research programs are turning to a new challenge: developing crops and technologies for farmers who do not irrigate their fields and who lack the income to purchase fertilizers and pesticides. The rice-breeding agenda at IRRI illustrates the shift in research priorities that will help meet their needs. In the sixties, the effort to raise yields of irrigated rice led to IR-8, IRRI's first widely planted high-yielding variety. When IR-8 began to experience serious pest infestation, breeders sought a wider variety of agronomic traits. IR-36 combined high yield with broad genetic resistance to pests, and it matured even more quickly than earlier varieties, permitting two crops to be harvested each year. IRRI's next successful rice strain, IR-64, was selected both for its broad resistance to pests and disease and for its more flavorful grain.[22]

In the eighties, breeders have further expanded their goals to developing rice varieties suited to adverse growing conditions—varieties that will be profitable for marginal and disadvantaged farmers. IRRI's breeding goals have evolved from a nearly exclusive emphasis on achieving peak yields with inputs of water and fertilizer to dependable production under a range of farming conditions.

In addition to appropriate crop varieties, poor farmers need alternative sources of plant nutrients. IRRI has begun to investigate opportunities to substitute farm-grown nutrient sources for purchased artificial fertilizers. Promising approaches for Asian farmers include the nitrogen-fixing blue-green algae sustained by a fern called *Azolla* that thrives in flooded rice paddies, and types of bacteria that could enhance soil fertility. Chinese farmers already use some of these methods quite successfully, and researchers in the Philippines have found that farmers who grew *Azolla* in their paddies were able to reduce their use of purchased fertilizer by 50 percent without lowering yields.[23]

Such innovations are not restricted to Asia. Scientists at the International Institute of Tropical Agriculture in Nigeria have identified a leguminous African shrub called *Sesbania* that may prove to be a low-cost nitrogen source for African rice farmers. Research in Colombia indicates that farmers can cut their needs for phosphate fertilizers in half by using certain fungi that help plant roots absorb phosphorus. IRRI recently created the International Biofertilizer Germplasm Conservation Center at its Philippines headquarters, where promising microbial sources of plant nutrients can be evaluated, stored, and distributed to researchers all over the world for testing.[24]

A range of other food crops is beginning to receive deserved research attention. Wheat and rice tend to be grown under relatively homogeneous conditions. Breeders of these crops drew on an enormous backlog of improved wheats and rices already available in Japan and North America, varieties whose pedigrees predated World War II. By contrast, improving the staple crops widely grown in Africa, and the potatoes, yams, and legumes grown throughout the Third World, is a much more challenging task. Such crops grow under widely divergent conditions, and have no comparable history of improvement. Systematic work on cassava and cowpeas in West Africa or potatoes in the Andes is little more than a decade old.

Efforts to raise the productivity of all staple crops in the years ahead depend on gathering a wide range of traditional varieties, crop relatives, and wild plants for breeding. Breeders need this genetic sampling to select the traits that strengthen resistance to pests and disease, and to tailor crops to grow under varied ecological conditions. Collecting and storing crop germplasm is coordinated by the International Board for Plant Genetic Resources (IBPGR). It is now a major responsibility of all the international centers. The value of distant crop relatives is likely to increase as biotechnology techniques are introduced that can speed up and simplify the tasks of breeding new varieties. IBPGR has initiated genetic resources programs in 50 countries, and national committees concerned with conservation of germplasm have been set up in over two dozen others.[25]

For most major food crops, germplasm collections of modern and traditional crop varieties are impressively broad. (See Table 3.) Except for wheat, however, scientists have not thoroughly investigated or collected the wild relatives of these crops. The unique genetic com-

Table 3: Estimated Number of Germplasm Samples Collected for Major Food Crops, and Coverage of Traditional Varieties and Wild Species

Crop	Samples in Major Genebanks	Estimated Share of Diversity Collected: Traditional Varieties	Estimated Share of Diversity Collected: Wild Species
	(thousands)	(percent)	
Wheat	400	95	60
Rice	200	70	10
Maize	70	90	—
Barley	250	40	10
Sorghum	90	80	10
Potato	42	95	—
Cowpea	18	75	1

Source: Adapted from Consultative Group on International Agricultural Research, "International Agricultural Research Centers: Achievements and Potential," unpublished draft, Washington D.C., August 1985.

binations of wild crop relatives are often lost as modern varieties and monocultures replace traditional farming methods. Such wild species may hold the key to improvements in the productivity of crops like sorghum and cowpeas that are crucial to Africa's food prospects.

Most of the world's food is supplied by a handful of crops selected by our neolithic ancestors. While farming technologies have advanced steadily, there have been few significant botanical innovations since the origins of agriculture. Most international research deals with just 16 widely grown crops, although at least 3,000 plants have been used for food at one time or another in history. Crops like teff, a hardy grass grown as a staple grain in Ethiopia, or amaranth, a grain and vegetable crop native to the Americas that is both nutritious and drought-tolerant, may prove better-suited than conventional crops to the environmental and economic conditions facing many Third World farmers.[26]

The network of international research centers may not be the wellspring of work on promising but unproven crops. By their charters, the centers are instructed to work on the most widely grown food crops. Research efforts focus on crops with proven potential and regions where the return to research investment is likely to be high. But restricting research to familiar crops may foreclose some important agricultural opportunities.

Naturalist Gary Nabhan, who has studied traditional food and medicinal plants native to the Sonoran Desert in the southwestern United States, believes that research on unconventional crops may be as valuable for insights on how to manage familiar crops as for novel agronomic possibilities. He writes, "By evaluating native desert plants as potential economic resources, and comparing them with conventional crops, we stand to learn something about the tradeoffs between short-term productivity and long-term persistence in unpredictable environments."[27]

Independent research centers have an important role to play in pursuing the agricultural opportunities that fall outside the mainstream of international research. The privately funded Rodale Research Center in Pennsylvania coordinates worldwide research on amaranth and

"In pursuit of higher productivity, many agricultural scientists overlooked the need for long-term sustainability."

maintains a germplasm collection containing 1,300 amaranth samples from Asia and Latin America. Scientists at Rodale and at the Land Institute in Kansas are investigating perennial grain polycultures as possible alternatives to today's annual corn and wheat monocultures, particularly for marginal lands. Agriculture based on perennials, though probably decades away, would offer several advantages over current practices, including reduced soil erosion, simplified weed control, improved water management, and enhanced soil fertility. Understanding perennial-based cropping practices could shed new light on how to reduce the environmental impact of more conventional farming practices.[28]

New crop varieties and technologies for farmers in developing countries will be essential in the years ahead. Biotechnologies may provide new generations of crop varieties to farmers left out of the green revolution. But to avoid the environmental and social costs associated with the last generation of agricultural technologies, tomorrow's innovations will have to be more consistent with regional agricultural traditions and better matched to the ecological context into which they are introduced.

Rediscovering Traditional Agriculture

Agricultural research has been needlessly hindered for two decades by pejorative attitudes toward traditional farming. Some scientists assumed that because peasant farmers produced low grain yields, their practices had little relevance to twentieth-century agriculture. Until recently, few researchers recognized the ecological and agronomic strengths of traditional practices that had allowed farmers over the centuries to maintain the land's fertility. In pursuit of higher productivity, many agricultural scientists overlooked the need for long-term sustainability.

Economic analysis reinforced the belief that traditional practices had little to offer in solving contemporary agricultural problems. In *Transforming Traditional Agriculture*, published in 1964, University of Chicago economist Theodore Schultz argued that peasant farmers were rational and efficient individuals who had reached the limits of their technologies. His conclusion: No significant increase in harvests could be achieved using only the resources and methods that traditional farmers had at their command. Schultz advocated investments in agricultural research, new technologies, and rural education that would allow traditional farmers to choose innovations to increase their productivity.[29]

Many scientists and policymakers, however, saw traditional methods as an obstacle to be eradicated rather than a basis for introducing new seeds and farming methods. The food crisis in India and throughout Asia in the late sixties lent a sense of urgency to efforts to promote the green revolution. The strengths of traditional practices and the reasons for their persistence were swept aside. A report by U.S. President Lyndon Johnson's Science Advisory Committee warned in 1966 that "the very fabric of traditional societies must be rewoven if the situation is to change permanently."[30]

Agricultural scientists have recently begun to recognize that many farming systems that have persisted for millennia exemplify careful management of soil, water, and nutrients; precisely the methods required to make high-input farming practices sustainable. This overdue reappraisal stems in part from the need to use inputs more efficiently, and in part from the growing interest in biological technologies. The complex challenge of Africa's food crisis in the early eighties forced scientists to look more closely at the methods used by peasant farmers. Many researchers today seek to "improve existing farming systems rather than attempting to transform them in a major way," according to William Liebhardt, director of research at the Rodale Research Center.[31]

Traditional farming systems face real agronomic limits, and can rarely compete ton for harvested ton with high-input modern methods. It is important to recognize these limitations, for they determine both how traditional practices can be modified and what such practices can contribute to the effort to raise agricultural productivity.

First, most traditional crop varieties have limited genetic potential for high grain yields. They are often large-leaved and tall, for example. These traits help farmers meet nonfood needs, supplying thatch,

"Agroforestry systems offer improvements in water-use efficiency and soil fertility that subsistence farmers can afford."

fuel, and fodder as well as food to farm households. Traditional varieties respond poorly to the two elements of agronomic management that make high grain yields possible: dense planting and artificial fertilizer. Despite these limitations, traditional varieties also contain genetic diversity that is invaluable to breeders in search of genes for disease- and pest-resistance and for other traits.[32]

Second, peasant farmers often have to plant in soils with serious nutrient deficiencies, where crop combinations and rotations are needed to help offset the limitations. Many tropical soils, for instance, lack sufficient nitrogen to sustain a robust crop. Soils in vast areas of semiarid Africa are deficient in phosphorus. High-yielding varieties, more efficient in converting available nutrients into edible grain, can rapidly deplete soil nutrients if they are planted in monocultures by peasant farmers who cannot afford to purchase supplemental fertilizers.[33]

Traditional agriculture, practiced under biological and physical limitations, often breaks down under growing population pressure. As rural populations grow, farmers try to squeeze more production from existing fields, accelerating the loss of fertility. Or they may cultivate new, marginal, or sloping land that is vulnerable to soil erosion and unsuited to farming.

Nonetheless, traditional methods can make an important contribution to efforts to raise agricultural productivity. They offer what Gerald Marten of the East-West Center in Hawaii calls "principles of permanence." They use few external inputs, accumulate and cycle natural nutrients effectively, protect soils, and rely on genetic diversity. "Neither modern Western agriculture nor indigenous traditional agriculture, in their present forms, are exactly what will be needed by most small-scale farmers," says Marten. "The challenge for agricultural research is to improve agriculture in ways that retain the strengths of traditional agriculture while meeting the needs of changing times."[34]

Farming methods like the traditional agroforestry systems of West Africa's Sahel region offer improvements in water-use efficiency and soil fertility that subsistence farmers can afford. Sahelian farmers traditionally planted their sorghum and millet crops in fields interspersed with a permanent intercrop of *Acacia albida* trees. Acacia trees fix nitrogen and improve the soil. In the Sahel, grain yields are often highest under an acacia's crown.[35]

Fields that include acacia trees produce more grain, support more livestock, and require shorter fallow periods between crops than fields sown to grain only. *Acacia albida* naturally enhances productivity by returning organic matter to the topsoil, drawing nutrients from deep soil layers to the surface, and changing soil texture so rainwater infiltrates the topsoil more readily. All of these benefits make farming on marginal lands more productive and profitable without requiring the farmer to purchase fertilizers year after year.[36]

Equally important, such improvements in soil structure, organic matter content, water-holding capacity, and biological nitrogen fixation allow the most productive application of conventional fertilizers. Programs promoting acacia-based agroforestry could complement fertilizer extension in semiarid countries—agroforestry playing a role analogous to irrigation. Governments that have modest fertilizer-promotion programs may find that they can maximize the benefits from fertilizer by promoting agroforestry as well.[37]

Legume-based crop rotations and traditional intercropping systems husband organic material and nutrients much more carefully than do modern monoculture practices. While organic manures and composts contribute significant amounts of nutrients in their own right, they can, like agroforestry, also magnify the contribution of small amounts of artificial fertilizers.

Research in Burkina Faso illustrates the complementary effect. (See Table 4.) This study looked at the contributions of straw, manure, and compost to sorghum yields with and without the addition of small amounts of artificial nitrogen. The results show that the most productive organic method, applying compost, can increase sorghum yields from 1.8 tons per hectare to 2.5 tons. Artificial fertilizer alone produced grain yields slightly higher than any of the organic practices. But the best result was achieved by combining compost with artificial fertilizer; this raised sorghum yields to 3.7 tons per hectare. The three

Table 4: Complementary Effect of Artificial and Organic Fertilizers on Sorghum Yields in Burkina Faso, 1981

Treatment[1]	Sorghum Yield: Without Artificial Nitrogen	Sorghum Yield: With 60 kg/ha Nitrogen
	(metric tons per hectare)	
No Organic Treatment	1.8	2.8
Sorghum Straw	1.6	3.4
Manure	2.4	3.6
Compost	2.5	3.7

[1]All organic materials applied at a rate of 10 tons per hectare.

Source: M. Sedogo, "Contribution à la valorisation des résidus culturaux en sol ferrugineux et sous climat semi-aride," doctoral thesis, Nancy, France, 1981, quoted in Herbert W. Ohm and Joseph G. Nagy, eds., *Appropriate Technologies for Farmers in Semi-Arid West Africa* (West Lafayette, Ind.: Purdue University International Programs in Agriculture, 1985).

organic practices increased the efficiency of nitrogen application by 20 to 30 percent. Given responsive crop varieties and small amounts of artificial fertilizer, traditional practices that cycle organic materials effectively would raise yields in the same manner.[38]

Some conventional analysts looking at the study would argue that fertilizer outperforms the organic practices. Yet exclusive reliance on fertilizer would sacrifice a significant part of the additional harvest. As French researcher Christian Pieri, who has worked in West Africa, points out, "Fertilization is a prime technique for increasing agricultural productivity in this part of the world, but in order to obtain a greater and lasting production it is indispensable to combine the effects of mineral fertilizers, the recycling of organic residues and biological nitrogen fixation, and also to optimize the use of local mineral resources such as natural phosphates."[39] Neglecting the local internal resources can undermine a farmer's investments in conventional inputs.

Intercropping, agroforestry, shifting cultivation, and other traditional farming methods mimic natural ecological processes, and the sustainability of many traditional practices lies in the ecological models they follow. This use of natural analogies suggests principles for the design of agricultural systems to make the most of sunlight, soil nutrients, and rainfall.

Shifting cultivation practices, such as bush-fallow methods in Africa, demonstrate how farmers can harness the land's natural regeneration. Farmers using bush-fallow systems clear fields by burning off the shrubs and trees. Ashes fertilize the first crop. After a couple of seasons, as nutrients are depleted, harvests begin to decline, so farmers abandon the field and move on to clear new land. Natural regeneration takes over; shrubs and trees gradually reseed the field, returning nutrients to the topsoil and restoring the land's inherent fertility. After 15 to 20 years, the land can be burned and cultivated again.[40]

The bush-fallow system has obvious limitations. It requires enormous amounts of land, and when population growth pushes farmers to return too quickly to abandoned fields, serious environmental deterioration can result. Declining land productivity in crowded countries like Rwanda is testimony to this danger. But even disintegrating systems offer a basis for designing productive and sustainable farming practices.

Researchers at the International Institute of Tropical Agriculture, for instance, have adapted the principles of natural regeneration in bush-fallow systems to a continuous-cultivation agroforestry system called alley cropping. Field crops are grown between rows of nitrogen-fixing trees; foliage from the trees enhances soil organic matter, while nitrogen fixed in root nodules increases soil fertility. A high level of crop production is possible without a fallow interval. Traditional shifting cultivation provided the model for this system.[41]

"The sustainability of many traditional practices lies in the ecological models they follow."

Conventional research tools can also be used to overcome the agronomic constraints that have limited traditional systems to low productivity. For decades, crop breeders have tailored varieties to respond to high levels of artificial fertilizers, assured water supplies, and dense monoculture plantings. Working with the genetic diversity available in traditional crop varieties, they can apply the same breeding methods to produce varieties better matched to the conditions faced by subsistence farmers. At an Agency for International Development workshop on regenerative farming practices, Charles Francis of the University of Nebraska concluded, "A new generation of varieties and hybrids adapted to marginal conditions and to intercropping could be the start of a new revolution aimed at meeting the needs of the majority of limited resource farmers in the developing world."[42]

Traditional practices exemplify efficiency and the regenerative approach to agricultural development. Yet until recently, a kind of myopia has kept the research community from recognizing the opportunities for agricultural innovations that lie in traditional practices. In West Africa, for example, 70 to 80 percent of the cultivated area is sown to combinations of crops in traditional intercropping systems. Cowpeas, one of Africa's most widely grown food staples, are always planted as an intercrop. But only about 20 percent of the research effort in sub-Saharan Africa focuses on intercropping.[43]

As the African examples described here show, researchers can use traditional principles to develop new techniques that preserve the land's stability and productivity even as populations increase. Though traditional methods have limitations, they are not archaic practices to be swept aside. Traditional farming constitutes a foundation upon which science can build.

Toward Appropriate Biotechnology

Most agricultural innovations of the past have been based on gradual refinements of technologies known at least since the Industrial Revolution and in some cases since the dawn of farming. But the 1953 discovery of the structure of DNA and the 1973 development of "recombinant DNA," or gene-splicing techniques, promise to change irretrievably the familiar landscape of agricultural development. Biotechnologies based on these insights allow scientists to identify the genes that control certain physical traits and to combine the genes of distantly related or unrelated plants and animals—two barriers that conventional plant breeders have never been able to overcome. Many analysts believe that agricultural applications of biotechnology will mark a watershed in the effort to raise productivity.

From 1920 to 1950, agriculture in industrial countries was dominated by mechanical technologies that dramatically increased the amount of food produced per worker and per hour. Shortly after World War II, the mechanical age gave way to the chemical age as farmers worldwide began to adopt artificial fertilizers and synthetic chemical pesticides, which vastly expanded their harvests per hectare. Biotechnologies shift the focus of research toward crop plants themselves. They have inaugurated a new era of agriculture likely to reshape research, development assistance, and farmers' choices. Biotechnologies may offer cheaper and quicker ways to improve Third World staples—including millet, cassava, and yams—than the costly innovations of the mechanical and chemical eras.[44]

Biotechnology encompasses an array of tools and applications that allow researchers to manipulate the genetic material of plants, microbes, and animals. These methods provide ways to modify the characteristics that are passed from one generation to the next. The vaccines, antibiotics, and reproductive technologies created through biotechnology and genetic engineering are already revolutionizing animal husbandry. Biotechnologies are not yet as widely applied to cultivated crops, in part because scientists understand less about plant genetics and physiology than about domestic animals.

Technical hurdles are not the only constraints on agricultural applications of biotechnology. So far, advances have been made in industrial countries, where public scrutiny is intense. The environmental risks posed by releasing gene-spliced microbes or plants into the environment remain poorly understood. Developing guidelines for the newly emerging technologies has led to a contentious public debate about genetic engineering. In the United States, debate has centered on proposals to release bacteria modified to retard the formation of frost on strawberry and potato plants. Because the bacteria could reproduce in the natural environment and thus spread beyond the fields where they were released, predicting environmental impacts is both more crucial and more complex a task than with many other technologies. Developing the "predictive ecology" that critics say is necessary for thorough environmental review, and enacting regulations that guard against the uncertainties will slow the marketing of commercial biotechnology products to industrial country farmers.[45]

The genetic engineering of plants is far more complex than modifying microbes, but it is also less controversial on environmental grounds. Crops with modified traits are under a farmer's direct control, and their reproduction and spread in the environment are both slower and more predictable. Crop characteristics like drought-tolerance, ability to withstand salty water, and pest resistance—the traits that have always concerned breeders—are a likely focus of the new technologies.

Thus, the major applications of biotechnologies to Third World crops will complement rather than replace conventional plant breeding. Developing new crop varieties can be an extraordinarily complex and time-consuming process. Identifying desirable characteristics, crossing parents, planting and growing the first generation of the cross, selecting the progeny that have the right mix of desired traits, and refining those characteristics through further breeding and screening can easily take a decade or longer. Conventional breeding of a new variety of wheat may involve thousands of carefully selected crosses.

By contrast, tissue culture, gene transfer, and other genetic techniques allow much of this work to be done in the laboratory, because researchers can manipulate single cells rather than entire plants. This saves space and time. Gene-splicing techniques allow researchers to transfer only specific traits into a crop. Such precision can help reduce the need to identify and eliminate full-grown plants carrying undesired genetic baggage—a problem when distantly related species or varieties are crossed.

Tissue culture techniques may revolutionize international gene banks by making it easier to store and manipulate crops that do not reproduce by setting seeds. These methods—which allow single plant cells to be sustained in laboratory flasks, multiplied, and regenerated into adult plants—are especially important for crops that propagate by roots or cuttings, such as cassava, potatoes, and yams. Tissue culture is also useful for propagating slow-growing species, including trees that hold promise for reforestation.[46]

Given the ability to modify virtually any plant characteristic and to tailor plants in precisely defined ways, biotechnology would seem to offer tools well-suited to agricultural development strategies that emphasize resource efficiency and farming's internal resources. According to the U.S. Office of Technology Assessment, "Most emerging technologies are expected to reduce substantially the land and water requirements for meeting future agricultural needs."[47] For example, it should eventually be possible to modify a plant's physiology to improve its efficiency in photosynthesis, enabling grains to produce more carbohydrate and thus, higher yields. The adaptations that allow some plants to lose very little water through their leaves in transpiration, transferred to more widely grown crops, could reduce irrigation needs. Developments like these could indeed diminish pressures on marginal lands and perhaps eliminate the need for costly capital investments in water supply projects.

There is nothing in the nature of biotechnologies that renders them inherently appropriate to a strategy of efficiency and regeneration, however. Many biotechnology innovations pose trade-offs rather than clearcut benefits. Although increasing photosynthetic efficiency could raise yields, it would likely lead to accelerated depletion of soil nutrients and heavier dependence on artifical fertilizers.

Another trade-off centers around herbicide resistance, a relatively uncomplicated genetic trait, which makes it an attractive research target. Researchers have already put considerable effort into developing crop plants that resist herbicides, allowing farmers to apply more of these chemicals. Much of this work is supported by the chemical companies that market herbicides.[48]

"The most significant factor that will affect the direction of agricultural biotechnology is the rapid shift of research to the private sector."

Herbicides have come to play a major role in industrial-country farming in recent years. High fuel costs and the need to conserve soil have prompted U.S. farmers to adopt reduced-tillage practices on 42 million hectares. These methods, which involve less plowing and leave topsoil covered with crop residues, employ herbicides rather than cultivation to control weeds. Conservation tillage is no longer restricted to industrial countries; scientists at the International Institute of Tropical Agriculture are also investigating more labor-intensive forms of these practices for small farmers to protect fragile tropical soils. In both industrial and developing countries, the soil- and energy-saving benefits of conservation tillage practices could be offset by the hazards of increased reliance on chemical herbicides.[49]

The most significant factor that will affect the direction of agricultural biotechnology is the rapid shift of research from the public to the private sector. This is especially evident in the United States. For nearly a century, public agricultural experiment stations and land grant universities sponsored by the U.S. Department of Agriculture (USDA) performed most agricultural research. Private seed companies often used the plant varieties developed by government-supported breeders. Over the last three decades, however, the private sector has assumed control of research efforts. Private companies now perform two-thirds of U.S. agricultural research.[50]

In biotechnology, the deck is stacked even further in favor of the private sector. USDA's Agricultural Research Service and Cooperative State Research Service support most public work in agricultural biotechnology, and these two federal programs spent less than $90 million on biotechnology research in 1984-85. Monsanto, which has the largest but by no means the only plant biotechnology research program among private U.S. corporations, has already invested $100 million in agricultural biotechnology development. Biotechnologies that affect agriculture in the years ahead will have a decidedly private-sector cast. With the exceptions of mechanization and the development of hybrid corn, that has not generally been true of important innovations in agriculture.[51]

Leaving research priorities to the marketplace may eclipse promising opportunities. Research efforts on crops will be proportional to the value of the crop and the size of the market. Because improving crops for small farmers in developing countries means producing low-cost agronomic innovations, many of which must be site-specific and thus not suitable for mass-marketing, crop improvement for the vast majority of the world's farmers offers little profit. Few private companies are likely to enter such an unpromising market. Consequently, investigations of minor crops like sorghum and millet, grown primarily by Third World subsistence farmers, will be neglected.

National research programs and the international research centers have an obvious stake in applying biotechnology. Refinements in plant breeding, technologies for germplasm storage and for plant evaluation and propagation, and new alternatives in pest control are exactly the kinds of innovations scientists need to extend research on developing-country food crops. It took decades of work to produce high-yielding varieties of wheat and rice. With biotechnology, comparable improvements in millet, sorghum, cassava, or tropical legumes could come more quickly.

The private-sector domination of biotechnology raises questions about the role new technologies will play in international research programs. Private companies may become competitors with the CGIAR-sponsored centers, particularly when it comes to improvements in major, widely traded crops like wheat and rice. The full exchange of scientific information that is essential to the international centers may be curtailed if it appears to compromise proprietary corporate research. Moreover, international centers may increasingly have to purchase or license new technologies that were formerly freely available through public channels. Finally, private firms will compete with the centers for scientific talent, and the centers may be unable to match the salaries, facilities, and security that corporate laboratories offer.[52]

Uncertainties cloud the prospects for national biotechnology programs as well. A few developing countries, notably Indonesia, the Philippines, and Thailand, have established national programs in agricultural biotechnology. The Philippines views its program as the first step toward an industrialization strategy based on biological materials that can help free the country from dependence on im-

"For Third World farmers who never shared in the agricultural advances of the green revolution, the issue is economic survival."

ported oil. Philippine scientists hope to use crop residues and by-products as raw materials to produce liquid fuels and industrial chemicals, and to develop food-processing industries with biotechnology methods. W.G. Padolina, of the National Institute of Biotechnology and Applied Microbiology at the University of the Philippines, writes, "The national strategy is to transform biomass biologically into food, fuel, fertilizers, and chemicals."[53]

Achieving these goals is certain to be costly. Few countries can afford the investment in equipment that major biotechnology programs entail, and some countries lack sufficient trained scientists to staff such programs. Agricultural biotechnology contrasts sharply in this regard with conventional plant-breeding programs, which require relatively modest capital investment.

Biotechnologies offer promising tools for more resource-efficient and sustainable agriculture. Technical hurdles must be overcome and environmental risks evaluated before that potential can be realized. But more troublesome from the standpoint of Third World agriculture is the degree to which the private sector will dominate agricultural biotechnologies. An expanded commitment to public research, at both the national and international levels, is needed to correct distortions of the research agenda and ensure that Third World priorities command attention. Public research in biotechnology consistent with resource-conserving and low-cost farming practices could counterbalance private-sector priorities.

Research for Sustainable Agriculture

The sense of urgency with which the green revolution was launched has largely disappeared from international agricultural development efforts. That several developing countries, formerly food importers, now have achieved food self-sufficiency has led some policymakers to question the value of assisting poor countries to increase food production further. But for Third World farmers who never shared in the agricultural advances of the green revolution, the issue is economic survival. Only by husbanding their scarce resources, regenerating their land, and raising their yields can these farmers improve their economic prospects. The reorientation of agricultural research and development assistance to meet their needs has begun, but deserves more attention and support.

An important bellwether of trends in international agricultural research is the funding of the world's 13 CGIAR-sponsored research centers. The budget grew from $21 million in 1972, when the system included just four centers, to over $100 million by 1980. This growth expanded the research mission to new crops and ecological zones. Spending increased more slowly to a level of about $170 million by the mid-eighties. While support for the centers remains strong, sufficient financial resources in the years ahead to underwrite more complex research tasks and changing technologies are by no means assured.[54]

CGIAR centers have established an important foundation of basic knowledge about staple food crops in the last 15 years. Opportunities to apply that knowledge could slip away if funding support stagnates. A large measure of responsibility for adapting crop research to local conditions rests with national research programs. Scientists at CGIAR hope that national programs will assume most of the responsibility for crop breeding in the years ahead. This would allow the international centers to focus on more "strategic" issues, including coordinating the conservation of crop genetic resources and applying biotechnology to staple crops.[55]

The international research agenda is shaped as much by new technologies as by critical agricultural needs. Biotechnology, now the principal focus of private-sector agricultural research, has captured the limelight; research administrators are scrambling to hire molecular biologists and redirect research programs. Taking this trend too far could be a serious mistake for public research institutions worldwide. As Cornell University sociologist Frederick Buttel counsels, "One must be cautious in assuming that there is only one scientific trajectory along which agricultural practices evolve."[56]

After supporting much of the work that led to the green revolution, the Rockefeller Foundation is now looking at ways to apply biotechnology to the crops overlooked by private-sector research. In 1983, the foundation redirected its program in agricultural sciences to emphasize biotechnology research on rice, the grain of least interest to private firms in industrial countries. In 1986, the foundation outlined a new agenda that included plans to extend biotechnology research to the improvement of sorghum, millet, and other neglected staple food crops—partly to counterbalance the private-sector emphasis on more widely grown commercial crops.[57]

The public research agenda can complement and compensate for the interests of the private sector in other ways as well. One way is to focus some portion of agricultural research on ecology. Robert Barker of Cornell University argues that public institutions like the U.S. land grant universities should shift their attention to the "development of the ecosystem sciences."[58] Agricultural technologies and practices that emphasize efficient use of resources and regenerative approaches are more likely to draw on the insights of ecology and evolutionary biology than on biochemistry.

In the past, responsibility for advances in resource efficiency and regenerative approaches has been left to independent institutions like the Rodale Research Center and some U.S. universities that have developed programs in agroecology. Participants in a 1980 Office of Technology Assessment workshop on biological technologies for agriculture noted that "much of the development of innovative technologies is occurring outside of, and perhaps in spite of, the national and international institutions normally considered responsible for maintaining natural resources and for dealing with problems of land quality and productivity."[59] That bias is slowly changing.

At the international level, the CGIAR centers have begun to acknowledge the importance of agricultural sustainability. The directors of the centers agreed in May 1986 to devote more research to raising crop productivity in ways that avoid environmental deterioration. The new emphasis on resource management goes beyond crop yield to encompass soil conservation, water management, and ways to help farmers reduce their reliance on purchased chemicals and fertilizers. In addition, the centers will work to develop technologies that can restore degraded croplands.[60]

The 230 million rural households in Africa, Asia, and Latin America that this research must reach are more isolated and face a far more complex set of agricultural constraints than their market-oriented counterparts. Actual conditions, rather than the ideal conditions in the experimental fields of research stations, determine the success or failure of new seeds, tools, or farming practices. Several international research centers have adopted a new approach to better understand the constraints faced by farmers on marginal lands. "Farming systems research" involves farmers and rural households directly in the research process. But how can the comparative handful of scientists in national and international research begin to reach a quarter of a billion households and refine technologies that match their individual circumstances? The answer must be a far more decentralized research effort that builds on farmer-scientist collaboration and equips farmers to produce innovations for themselves.[61]

The reappraisal of traditional practices is a step toward this collaboration. According to Paul Richards of University College in London who has worked with Nigerian farmers, indigenous agricultural knowledge is "the single largest knowledge resource not yet mobilized in the development enterprise." In his book, *Indigenous Agricultural Revolution*, Richards documents how traditional farmers in West Africa have modified farming practices on the basis of carefully controlled experiments, ranging from selection of rice varieties to the control of grasshoppers. He suggests that mainstream researchers have as much to learn from the partnership with small farmers as the farmers themselves.[62]

The challenge for agricultural research at all levels is no longer a problem of one-way "technology transfer," as so many people perceived the green revolution. Innovations and insights that help raise agricultural productivity will flow in both directions—between researchers and farmers, between developing and industrial countries. Success in the low-productivity fields of the Third World can suggest new ways of managing agricultural resources that farmers in Iowa or France could employ as well.

The conservation and use of crop genetic diversity illustrates the international convergence of interests in raising agricultural productivity in the Third World. The tools of biotechnology are needed to store, evaluate, and manipulate the genes in traditional crop varieties and wild plants needed for crop breeding. Yet much of the diversity itself still resides in farmers' fields, where crops are adapted to the idiosyncracies of local rainfall, soils, and cultivation methods. "Neither money, talent and technology, nor unimproved germplasm alone can create improved crop plants—the former must be applied to the latter," points out Steven Witt in his book *Biotechnology and Genetic Diversity*. "And that means cooperation between those who have the talent and technology and those with the necessary germplasm."[63]

The world is far from having solved the problems of agricultural productivity. The conventional approach to raising productivity—combining new crop varieties with fertilizers, pesticides, and heavy use of energy—succeeded dramatically in increasing food production in industrial countries and in parts of the Third World. But new approaches are needed to reach farmers who could not afford the conventional technologies, as well as to correct inequities in the distribution of resources and confront widespread environmental problems. Complementing the use of conventional resources with innovative biological technologies that maximize agriculture's internal resources can ensure the affordable and sustainable gains in agricultural productivity that the world needs in the years ahead.

Notes

1. Data on adoption of high-yielding varieties are from Dana G. Dalrymple, *Development and Spread of High-Yielding Rice Varieties in Developing Countries* (Washington, D.C.: U.S. Agency for International Development, 1986) and Dana G. Dalrymple, *Development and Spread of High-Yielding Wheat Varieties in Developing Countries* (Washington, D.C.: U.S. Agency for International Development, 1986). Regional grain area data are from U.S. Department of Agriculture (USDA), Economic Research Service (ERS), "World Indices of Agricultural and Food Production 1950-85," unpublished printout, Washington, D.C., April 1986.

2. Population data are based on estimates of agriculturally active populations from Food and Agriculture Organization, *1984 Production Yearbook* (Rome: 1985).

3. Population projections are from Population Reference Bureau, *1985 World Population Data Sheet* (Washington, D.C.: 1985). Grain-yield projections are by Worldwatch Institute, based on world grain utilization data from USDA, Foreign Agricultural Service, *Foreign Agriculture Circular—Grains,* FG-9-86 (Washington, D.C.: 1986).

4. For descriptions of traditional farming, see D.B. Grigg, *The Agricultural Systems of the World* (Cambridge, England: Cambridge University Press, 1974); Gerald G. Marten, ed., *Traditional Agriculture in Southeast Asia* (Boulder, Colo.: Westview Press, 1986); and U.S. Office of Technology Assessment (OTA), *Africa Tomorrow: Issues in Technology, Agriculture, and U.S. Foreign Aid* (Washington, D.C.: U.S. Government Printing Office, 1984).

5. USDA, ERS, "World Indices."

6. World Bank, *China: Agriculture to the Year 2000* (Washington, D.C.: 1985); Bruce Stone, "Chinese Fertilizer Application in the 1980s and 1990s: Issues of Growth, Balance, Allocation, Efficiency and Response," in U.S. Congress, Joint Economic Committee, *China's Economy in the Eighties* (Washington, D.C.: U.S. Government Printing Office, forthcoming).

7. Stone, "Chinese Fertilizer Application in the 1980s and 1990s."

8. Organisation for Economic Co-operation and Development, *The State of the Environment 1985* (Paris: 1985).

9. Charles A. Francis and Richard R. Harwood, *Enough Food: Achieving Food Security through Regenerative Agriculture* (Emmaus, Penn.: Rodale Institute, 1985).

10. Robert Rodale, "Internal Resources and External Inputs—The Two Sources of All Production Needs," in Rodale Institute, *Regenerative Farming Systems* (Emmaus, Penn.: 1985).

11. D.F. Bezdicek, R.F. Mulford, and B.H. Magee, "Influence of Organic Nitrogen on Soil Nitrogen, Nodulation, Nitrogen Fixation, and Yield of Soybeans," *Soil Science Society of America Proceedings,* March-April 1974; D.F. Bezdicek, "Biotechnology and Farming Systems: On-Farm Applications and Consequences," in Institute for Alternative Agriculture, *Biotechnology and Agriculture: Implications for Sustainability* (Greenbelt, Md.: 1986); and D. F. Bezdicek, Washington State University, Pullman, Wash., private communication, August 6, 1986.

12. For a review of regenerative practices, see Francis and Harwood, *Enough Food;* Rodale Institute, *Regenerative Farming Systems;* and *Proceedings of Workshop on Resource-Efficient Farming Methods for Tanzania* (Emmaus, Penn.: Rodale Press, 1983).

13. Dana G. Dalrymple, "The Development and Adoption of High-Yielding Varieties of Wheat and Rice in Developing Countries," *American Journal of Agricultural Economics,* December 1985.

14. Figures compiled from Dalrymple, *Development and Spread of High-Yielding Rice Varieties;* Dalrymple, "The Development and Adoption of High-Yielding Varieties"; Inter-American Development Bank, *Economic and Social Progess in Latin America: 1986 Report* (Washington, D.C.: 1986).

15. Consultative Group on International Agricultural Research (CGIAR), *1984 Annual Report* (Washington, D.C.: 1985); *News from the CGIAR,* January-April 1986; Michael Lipton with Richard Longhurst, *Modern Varieties, International Agricultural Research, and the Poor* (Washington, D.C.: World Bank, 1985).

16. Dalrymple, "The Development and Adoption of High-Yielding Varieties"; USDA, ERS, "World Indices"; Dunstan S.C. Spencer, "Agricultural Research: Lessons of the Past, Strategies for the Future," in Robert J. Berg and Jennifer Seymour Whittaker, eds., *Strategies for African Development* (Berkeley: University of California Press, 1986).

17. CGIAR, *1984 Annual Report.*

18. Lipton with Longhurst, *Modern Varieties, International Agricultural Research, and the Poor.*

19. Ibid. Estimate of area and population unaffected by new agricultural technologies is by Worldwatch Institute based on Dalrymple, *Development and Spread of High-Yielding Wheat Varieties;* USDA, "World Indices"; and Food and Agriculture Organization, *Production Yearbook.*

20. International Development Research Center, *The Fragile Web: The International Agricultural Research System* (Ottawa: 1983); CGIAR, *1984 Annual Report.*

21. Donald L. Plucknett and Nigel J.H. Smith, "Sustaining Agricultural Yields," *BioScience,* January 1986.

22. International Rice Research Institute (IRRI), *IRRI Highlights 1985: Accomplishments and Challenges* (Manila, Philippines: 1986).

23. Ibid; CGIAR, *Summary of International Agriculture Research Centers: A Study of Achievements and Potential* (Washington, D.C.: 1985).

24. International Institute of Tropical Agriculture (IITA), *IITA Annual Report and Research Highlights 1985* (Ibadan, Nigeria: 1986); "Mycorrhizae: Can Africa Benefit?," *International Livestock Center for Africa Newsletter* (Addis Ababa, Ethiopia), July 1986; IRRI, *IRRI Highlights 1985.*

25. J.G. Hawkes, *Plant Genetic Resources: The Impact of the International Agricultural Research Centers* (Washington, D.C.: World Bank, 1985).

26. Debora MacKenzie, "Ethiopia: Famine amid Genetic Plenty," *New Scientist,* August 8, 1985; Jonathan B. Tucker, "Amaranth: The Once and Future Crop," *BioScience,* January 1986.

27. Gary Paul Nabhan, *Gathering the Desert* (Tucson: The University of Arizona Press, 1985).

28. Michael Philips, "Rodale Research Center Holds Premier Amaranth Collection," *Diversity,* No. 9, 1986; William Liebhardt and Charles S. Kauffman, Rodale Research Center, Kutztown, Penn., private communications, March 26, 1986; Wes Jackson, *New Roots for Agriculture* (San Francisco: Friends of the Earth, 1980).

29. Theodore Schultz, *Transforming Traditional Agriculture* (New Haven, Conn.: Yale Univeristy Press, 1964).

30. Quoted in Sterling Wortman and Ralph W. Cummings, Jr., *To Feed This World* (Baltimore, Md.: The Johns Hopkins University Press, 1978).

31. W.C. Liebhardt, C.A. Francis, and M. Sands, "Research Needs for the Development of Resource Efficient Technologies," in Rodale Institute, *Regenerative Farming Systems.*

32. Peter R. Jennings, "The Amplification of Agricultural Production," *Scientific American,* September 1976.

33. Ibid.

34. Gerald G. Marten, "Traditional Agriculture and Agricultural Research in Southeast Asia," in Marten, ed., *Traditional Agriculture in Southeast Asia.*

35. Traditional acacia-based systems are described in National Research Council, Board on Science and Technology for International Development, *Environmental Change in the West African Sahel* (Washington, D.C.: National Academy Press, 1983).

36. Ibid.

37. Michael McGahuey, *Impact of Forestry Initiatives in the Sahel* (Washington, D.C.: Chemonics, 1986).

38. Christian Pieri, "Food Crop Fertilization and Soil Fertility: The IRAT Experience," in Herbert W. Ohm and Joseph G. Nagy, eds., *Appropriate Technologies for Farmers in Semi-Arid West Africa* (West Lafayette, Ind.: Purdue University International Programs in Agriculture, 1985).

39. Ibid.

40. Shifting cultivation is described in Grigg, *Agricultural Systems.*

41. Current research in alley cropping is described in IITA, *IITA Annual Report 1985.*

42. C.A. Francis et al., "Resource Efficient Farming Systems and Technologies," in Rodale Institute, *Regenerative Farming Systems.*

43. Spencer, "Agricultural Research."

44. OTA, *Technology, Public Policy, and the Changing Structure of American Agriculture* (Washington, D.C.: U.S. Government Printing Office, 1986).

45. For overviews of the environmental implications of biotechnologies, see Jack Doyle, "Biotechnology Research and Agricultural Stability," *Issues in Science and Technology,* Fall 1985; and Jack Doyle, *Altered Harvest* (New York: Viking Penguin Inc., 1985).

46. CGIAR, *1984 Annual Report.*

47. OTA, *Technology, Public Policy, and the Changing Structure of American Agriculture.*

48. Marjorie Sun, "Engineering Crops to Resist Weed Killers," *Science,* March 21, 1986.

49. Data on conservation tillage in the United States are from *No-Till Farmer,* March 1986. African conservation tillage research is described in IITA, *Tasks for the Eighties: A Long-Range Plan* (Ibadan, Nigeria: 1981), and IITA, *IITA Annual Report 1984* (Ibadan, Nigeria: 1985).

50. Frederick H. Buttel et al., "Genetic Engineering and the Restructuring of Agricultural Research," *The Rural Sociologist,* Vol. 3, No. 3, 1983; the history of public agricultural research in the United States is reviewed in Yujiro Hayami and Vernon W. Ruttan, *Agricultural Development* (Baltimore, Md.: The Johns Hopkins University Press, 1985).

51. Estimate of USDA-supported biotechnology research is from U.S. General Accounting Office, *Biotechnology: The U.S. Department of Agriculture's Biotechnology Research Efforts* (Washington, D.C.: 1985); Frederick Buttel, "Biotechnology and Alternative Agriculture: An Overview of the Major Issues and Concerns," in Institute for Alternative Agriculture, *Biotechnology in Agriculture.*

52. These and other challenges facing the international research centers are discussed in Frederick H. Buttel and Randolph Barker, "Emerging Agricultural Technologies, Public Policy, and Implications for Third World Agriculture: The Case of Biotechnology," *American Journal of Agricultural Economics,* December 1985; and in F.H. Buttel, B. Kenney, and J. Kloppenburg, Jr., "The IARCs and the Development and Application of Biotechnologies in Developing Countries," in IRRI, *Biotechnology in International Agricultural Research* (Manila: 1985).

53. W.G. Padolina, "Strategies to Develop Biotechnology in the Philippines," in IRRI, *Biotechnology in International Agricultural Research.*

54. CGIAR, *1984 Annual Report.*

55. *News from the CGIAR,* May-August 1986.

56. Buttel, "Genetic Engineering and the Restructuring of Agricultural Research."

57. Randolph Barker, "Biotechnology and Farming Systems: An International Perspective," in Institute for Alternative Agriculture, *Biotechnology in Agriculture;* The Rockefeller Foundation, *The President's Review and Annual Report 1983* (New York: 1983); Kathleen Teltsch, "Rockefeller Unit Doubles Its Third-World Aid," *New York Times,* May 4, 1986; The Rockefeller Foundation, "The Rockefeller Foundation in the Developing World," New York, May 1986.

58. Robert Barker, "The Changed World of Research Opportunities," in Martin Gibbs and Carla Carlson, eds., *Crop Productivity—Research Imperatives Revisited,* an international conference held at Boyne Highlands Inn, Boyne, Michigan, October 13-18, 1985, and Airlie House, Virginia, December 11-13, 1985.

59. OTA, *Innovative Biological Technologies for Lesser Developed Countries* (Washington, D.C.: U.S. Government Printing Office, 1985).

60. *News from the CGIAR,* May-August 1986.

61. For several perspectives on farming systems research, see Joyce Lewinger Moock, ed., *Understanding Africa's Rural Households and Farming Systems* (Boulder, Colo.: Westview Press, 1986); the potential for farmer-scientist collaboration is described in Marten, "Traditional Agriculture and Agricultural Research in Southeast Asia."

62. Paul Richards, *Indigenous Agricultural Revolution* (London: Hutchinson & Co., Ltd., 1985).

63. Steven C. Witt, *Biotechnology and Genetic Diversity* (San Francisco: California Agricultural Lands Project, 1985).

EDWARD C. WOLF is a Senior Researcher with Worldwatch Institute. His research deals with agricultural development and the management of biological resources. He is a graduate of Williams College, where he studied biology and participated in the Environmental Studies Program.

THE WORLDWATCH PAPER SERIES

No. of Copies*

- ______ 1. **The Other Energy Crisis: Firewood** by Erik Eckholm.
- ______ 2. **The Politics and Responsibility of the North American Breadbasket** by Lester R. Brown.
- ______ 3. **Women in Politics: A Global Review** by Kathleen Newland.
- ______ 4. **Energy: The Case for Conservation** by Denis Hayes.
- ______ 5. **Twenty-two Dimensions of the Population Problem** by Lester R. Brown, Patricia L. McGrath, and Bruce Stokes.
- ______ 6. **Nuclear Power: The Fifth Horseman** by Denis Hayes.
- ______ 7. **The Unfinished Assignment: Equal Education for Women** by Patricia L. McGrath.
- ______ 8. **World Population Trends: Signs of Hope, Signs of Stress** by Lester R. Brown.
- ______ 9. **The Two Faces of Malnutrition** by Erik Eckholm and Frank Record.
- ______ 10. **Health: The Family Planning Factor** by Erik Eckholm and Kathleen Newland.
- ______ 12. **Filling The Family Planning Gap** by Bruce Stokes.
- ______ 13. **Spreading Deserts—The Hand of Man** by Erik Eckholm and Lester R. Brown.
- ______ 14. **Redefining National Security** by Lester R. Brown.
- ______ 15. **Energy for Development: Third World Options** by Denis Hayes.
- ______ 16. **Women and Population Growth: Choice Beyond Childbearing** by Kathleen Newland.
- ______ 17. **Local Responses to Global Problems: A Key to Meeting Basic Human Needs** by Bruce Stokes.
- ______ 18. **Cutting Tobacco's Toll** by Erik Eckholm.
- ______ 19. **The Solar Energy Timetable** by Denis Hayes.
- ______ 20. **The Global Economic Prospect: New Sources of Economic Stress** by Lester R. Brown.
- ______ 21. **Soft Technologies, Hard Choices** by Colin Norman.
- ______ 22. **Disappearing Species: The Social Challenge** by Erik Eckholm.
- ______ 24. **The Worldwide Loss of Cropland** by Lester R. Brown.
- ______ 25. **Worker Participation—Productivity and the Quality of Work Life** by Bruce Stokes.
- ______ 26. **Planting for the Future: Forestry for Human Needs** by Erik Eckholm.
- ______ 27. **Pollution: The Neglected Dimensions** by Denis Hayes.
- ______ 28. **Global Employment and Economic Justice: The Policy Challenge** by Kathleen Newland.
- ______ 29. **Resource Trends and Population Policy: A Time for Reassessment** by Lester R. Brown.
- ______ 30. **The Dispossessed of the Earth: Land Reform and Sustainable Development** by Erik Eckholm.
- ______ 31. **Knowledge and Power: The Global Research and Development Budget** by Colin Norman.
- ______ 32. **The Future of the Automobile in an Oil-Short World** by Lester R. Brown, Christopher Flavin, and Colin Norman.
- ______ 33. **International Migration: The Search for Work** by Kathleen Newland.
- ______ 34. **Inflation: The Rising Cost of Living on a Small Planet** by Robert Fuller.
- ______ 35. **Food or Fuel: New Competition for the World's Cropland** by Lester R. Brown.
- ______ 36. **The Future of Synthetic Materials: The Petroleum Connection** by Christopher Flavin.
- ______ 37. **Women, Men, and The Division of Labor** by Kathleen Newland.
- ______ 38. **City Limits: Emerging Constraints on Urban Growth** by Kathleen Newland.
- ______ 39. **Microelectronics at Work: Productivity and Jobs in the World Economy** by Colin Norman.

*Worldwatch Papers 11 and 23 are out of print.

_____ 40. **Energy and Architecture: The Solar and Conservation Potential** by Christopher Flavin.
_____ 41. **Men and Family Planning** by Bruce Stokes.
_____ 42. **Wood: An Ancient Fuel with a New Future** by Nigel Smith.
_____ 43. **Refugees: The New International Politics of Displacement** by Kathleen Newland.
_____ 44. **Rivers of Energy: The Hydropower Potential** by Daniel Deudney.
_____ 45. **Wind Power: A Turning Point** by Christopher Flavin.
_____ 46. **Global Housing Prospects: The Resource Constraints** by Bruce Stokes.
_____ 47. **Infant Mortality and the Health of Societies** by Kathleen Newland.
_____ 48. **Six Steps to a Sustainable Society** by Lester R. Brown and Pamela Shaw.
_____ 49. **Productivity: The New Economic Context** by Kathleen Newland.
_____ 50. **Space: The High Frontier in Perspective** by Daniel Deudney.
_____ 51. **U.S. and Soviet Agriculture: The Shifting Balance of Power** by Lester R. Brown.
_____ 52. **Electricity from Sunlight: The Future of Photovoltaics** by Christopher Flavin.
_____ 53. **Population Policies for a New Economic Era** by Lester R. Brown.
_____ 54. **Promoting Population Stabilization** by Judith Jacobsen.
_____ 55. **Whole Earth Security: A Geopolitics of Peace** by Daniel Deudney.
_____ 56. **Materials Recycling: The Virtue of Necessity** by William U. Chandler.
_____ 57. **Nuclear Power: The Market Test** by Christopher Flavin.
_____ 58. **Air Pollution, Acid Rain, and the Future of Forests** by Sandra Postel.
_____ 59. **Improving World Health: A Least Cost Strategy** William U. Chandler
_____ 60. **Soil Erosion: Quiet Crisis in the World Economy** by Lester Brown and Edward Wolf.
_____ 61. **Electricity's Future: The Shift to Efficiency and Small-Scale Power** by Christopher Flavin.
_____ 62. **Water: Rethinking Management in an Age of Scarcity** by Sandra Postel.
_____ 63. **Energy Productivity: Key to Environmental Protection and Economic Progress** by William U. Chandler.
_____ 64. **Investing in Children** by William U. Chandler.
_____ 65. **Reversing Africa's Decline** by Lester Brown and Edward Wolf.
_____ 66. **World Oil: Coping With the Dangers of Success** by C. Flavin.
_____ 67. **Conserving Water: The Untapped Alternative** by Sandra Postel
_____ 68. **Banishing Tobacco** by William U. Chandler
_____ 69. **Decommissioning: Nuclear Power's Missing Link** by Cynthia Pollock
_____ 70. **Electricity For A Developing World** by Christopher Flavin
_____ 71. **Altering the Earth's Chemistry: Assessing the Risks** by Sandra Postel
_____ 72. **The Changing Role of the Market in National Economies** by William U. Chandler.
_____ 73. **Beyond the Green Revolution: New Approaches for Third World Agriculture** by Edward C. Wolf.
_____ **Total Copies**

Bulk Copies (any combination of titles)
2-5: $3.00 each 6-20: $2.00 each 21 or more: $1.00 each

Calendar Year Subscription (1986 subscription begins wth Paper 68)
U.S. $25.00 _____

Make check payable to Worldwatch Institute
1776 Massachusetts Avenue NW, Washington, D.C. 20036 USA

Enclosed is my check for U.S. $ _____

name

address

city **state** **zip/country**

CHOPIN'S
MUSICAL STYLE

CHOPIN'S MUSICAL STYLE

BY

GERALD ABRAHAM

GREENWOOD PRESS, PUBLISHERS
WESTPORT, CONNECTICUT

Library of Congress Cataloging in Publication Data

Abraham, Gerald Ernest Heal, 1904-
Chopin's musical style.

Includes index.
Reprint of the ed. published by Oxford University Press, London, New York.
1. Chopin, Fryderyk Franciszek, 1810-1849. Works.
I. Title.
ML410.C54A4 1980 786.1'092'4 79-25521
ISBN 0-313-22251-7 lib. bdg.

This is a reprint of the 1968 impression of the 1939 first edition

This reprint has been authorized by the Oxford University Press

Reprinted in 1980 by Greenwood Press, a division of Congressional Information Service, Inc., 51 Riverside Avenue, Westport, Connecticut 06880

Printed in the United States of America

10 9 8 7 6 5 4 3 2 1

CONTENTS

INTRODUCTION

THIS little book is an attempt to show the gradual unfolding and maturing of Chopin's musical mind, to investigate the working of that mind, and to isolate and define those qualities which make Chopin's work unique in the history of music. Every one admits Chopin's charm (though some may find it not to their taste), but it is probable that comparatively few understand precisely how he worked his miracles.

But, it may be asked, is it of any use to the practical musician, particularly to the pianist, to know this? Emphatically, yes. It is becoming more and more generally recognized that intelligent interpretation is practically impossible without thorough knowledge of a composer's style. That is true of all composers in some degree, and particularly true of Chopin. To take one example: more than sixty years ago Sir Henry Hadow[1] wrote: 'In the twelfth bar of the well-known Nocturne in E flat (Op. 9, No. 2), there is a connecting passage which, when we see it on paper, seems to consist of a rapid series of remote and recondite modulations. *When we hear it played in the manner which Chopin intended* [my italics] we feel that there is only one real modulation, and that the rest of the passage is an iridescent play of colour, an effect of superficies, not an effect of substance.' Without real understanding of

[1] *Studies in Modern Music* (Second Series). Seeley, Service & Co., Ltd., 1894.

Chopin's style, it is impossible to distinguish between 'effects of superficies' and 'effects of substance', and hence to play them as they were intended to be played. To put it in another way: the passage is a harmonic parenthesis—and the parenthetic effect must be conveyed by the player.

And such parentheses occur not only in the harmony, but in the formal structure. Writing of the E major Scherzo, Op. 54, Hugo Leichtentritt[1] draws attention to the eight-bar 'asides' (bars 25–32, 57–64, etc.) which his analysis shows to be inserted in the main argument, and which 'should be different in tone-colour from the real melody-bearing matter. They should be played incidentally as it were—less emphatically than the main melodic line'. Intelligent analysis is one of the interpreter's most valuable tools.

This book should also be of help to the student of harmony. For, as is now generally admitted, Chopin's chromaticism marks a stage of the greatest importance in the evolution of the harmonic language. He is the forerunner of Liszt and Wagner, and hence of the modern atonalists, the first composer seriously to undermine the solid system of diatonic tonalism created by the Viennese classical masters and their contemporaries in other countries.

Writing for the practical musician, I have left almost untouched two problems of great interest to the musicologist: the origins of Chopin's musical style, and the true chronological order of his compositions.

[1] *Analyse der Chopin'sche Klavierwerke*. Berlin. 1922.

It is fairly obvious that the young Chopin was musically descended from Hummel and Field and Moscheles rather than from Mozart, whom he adored, and Beethoven, whom he seems to have known imperfectly and admired but moderately. There are traces here and there of a slight Weberian strain. But what of Schubert? 'I think Chopin knew too little of his music to be appreciably influenced by him', says Niecks.[1] 'At any rate, I fail to perceive how and where the influence reveals itself.' Yet they have certain traits in common, though whether Chopin acquired them from Schubert or from the circumambient musical atmosphere is a matter for debate—or, rather, for investigation. Again, there is the influence of Spohr, whom we know Chopin did admire (see his letters of October 3, 1829, and September 18, 1830). Was Chopin's chromatic harmony influenced by Spohr's? And if so, to what extent?

And what of the influence of Joseph Elsner (1769–1854), the Silesian-born musician who had the honour of being Chopin's only composition teacher? Elsner's innumerable compositions in every genre, from opera and symphony to the small piano piece, are generally inaccessible. But, according to Niecks, 'most accounts agree in stating that he wrote in the style of the modern Italians, that is to say, what were called the modern Italians in the later part of the last and the earlier part of this century'. 'Elsner's productions are in the style

[1] *Frederick Chopin as a Man and Musician.* Novello. 1888. (Third edition, 1902.)

of Paër and Mayer's music', says Fétis. According to Dr. Maria Ottich, his twenty-eight polonaises 'anticipate in certain traits the great perfector of this form'. And the author of the article 'Elsner' in Schilling's *Universal-Lexikon der Tonkunst*[1] remarks that 'one forgives him readily [in consideration of the general excellence of his style] the offences against the law of harmonic connexion that occur here and there, and the facility with which he sometimes disregards the fixed rules of strict part-writing'. It is inconceivable that Chopin was quite uninfluenced by his master. But how great was the influence? In what direction? Are Chopin's own 'offences against the law of harmonic connexion' (as then administered) reflections of Elsner's?

These are questions which I have not attempted to answer, though to do so would be a fascinating task. But from the point of view of the working musician of to-day the music of Hummel and Spohr and Moscheles—to say nothing of Elsner—is unhappily as dead as the men who wrote it, and Field's nearly so, while Chopin's is gloriously alive. Moreover, no matter how much Chopin may have been indebted to these others in his caterpillar-and-chrysalis stage, by the time he emerged at twenty-one as the lovely chatoyant butterfly we think of as the true Chopin, he was as free from debt to predecessors as any composer in the whole history of music. I have confined myself here to tracing the evolution of the butterfly from the caterpillar without bothering unduly about the

[1] Quoted by Niecks.

origin of the eggs from which the caterpillar emerged.

I have concernèd myself with chronology only so far as to divide Chopin's output into three main periods: the positively and comparatively immature works written before his arrival in Paris in September, 1831; the works of 1831–40; and the compositions of the last nine years. (A number of works have been re-dated in the light of Mr. Arthur Hedley's list in the fifth edition of *Grove*.)

This practice of drawing lines at the end of a creative artist's 'periods' is as unavoidable as it is detestable.[1] And in this case the lines are drawn less arbitrarily than may appear at first sight. No one will deny that the boyish compositions of 1822–8 are immature. And though, as Niecks says, 'at the very first glance it becomes evident that the works of the last two years (1829–31) are decidedly superior to those he wrote before that time', they hardly belong—except the first set of Etudes, Op. 10—to the true Chopin canon. Chopin did not fully become himself till his arrival in Paris.

The division between the works of the second and third periods is much less clearly defined and less easy to justify. Nevertheless, while many of the third-period compositions are hardly distinguish-

[1] 'It would be easy to classify the Polish master's works according to three or even four successive styles, but I have no taste for this cheap kind of useless ingenuity', says Niecks. But such classification if cheap is certainly not useless; on the contrary, it is a valuable convenience.

able from those of the second, I hope to be able to show that in his last nine years Chopin produced a number of works more powerfully conceived, more organically constructed (at the risk of being misunderstood, I am tempted to say 'more symphonic'), than any he had written before. They also show a new tendency to polyphony: and a slight—a very slight—pretext for drawing the line just here is provided by a sentence in a letter to Fontana, undated but apparently written in July or August, 1841: 'Send without fail Cherubini's *traité*; I think it's *du contrepoint* (I don't remember the title well)' (the *Cours de Contrepoint et de la Fugue*, which had been published in 1835).

In my choice of music-type examples, I have worked on the assumption that the reader possesses copies of Chopin's better-known works, which he would be well advised to keep at his side. In the case of lesser-known compositions, I have used music-type more freely to facilitate reference.

I

EVOLUTION OF A MUSICAL PERSONALITY

(1822–1831)

THE more important compositions of Chopin's nonage may be divided into two main groups: more or less showy concert-pieces in forms which he abandoned in later life:

Rondeau in C minor, Op. 1 (1825)
Variations sur un air national allemand (Op. posth.) (1826)
Rondeau à la mazurka, Op. 5 (1826)
'La ci darem la mano' de l'opera *Don Juan.* Varie, Op. 2 (1827)[1]
Rondeau pour deux pianos, Op. 73 (1828)
Fantaisie sur des airs polonais, Op. 13 (1828)
Krakowiak, Op. 14 (1828)

and the first essays in those smaller forms which he afterwards practised so sucessfully. With these we may group the youthful C minor Piano Sonata:

Mazurkas in G major and B flat major (1825), in A minor, Op. 68, No. 2 (1827), in D major (1829), in C and F, Op. 68, Nos. 1 and 3 (1829), and Op 6 and 7 (mostly 1830–1)

[1] Not 1828, as Niecks supposed. Chopin mentions the work in letter written towards the end of 1827.

Nocturnes in E minor, Op. 72, No. 1 (1827), Op. 9, and Op. 15, Nos. 1 and 2 (1830–1)

Polonaises in G sharp minor (1822), D minor, B flat major and F minor, Op. 71, Nos. 1, 2 and 3 (1827–29)

Sonata in C minor, Op. 4 (1827)

Valses in B minor, Op. 69, No. 2, in D flat, Op. 70, No. 3, and in E minor (all about 1829)

In addition there are the quite unimportant Marche funèbre in C minor, Op. 72, No. 2 (1829), and the Trois Ecossaises (Op. 72, No. 3) (1826).

To these two groups we must add a third consisting of the most significant 'border-line' works written during the transition period 1829–31, all important, yet not quite to be classed with the works of maturity:

Concerto in F minor, Op. 21 (1829)
Concerto in E minor, Op. 11 (1830)
Grand Polonaise brillante, Op. 22 (1830–1)
Douze Grandes Etudes, Op. 10 (1829–32)

Consideration of this third group, however, had better be postponed till we have looked a little more closely at the first two.

It will be seen that the earliest generally known work of Chopin's is the Polonaise in G sharp minor (No. 14 of the Polish National Edition), the composition of which is commonly attributed to the year 1822, when the composer was twelve. Niecks considered that 'on account of the *savoir faire* and invention exhibited in it' this must have been written at 'a considerably later time'. But actually there is little in the work that might not have come from a very talented boy of twelve. Neither the spirit of the piece nor its harmony or structure distinguishes it from hundreds of contemporary brilliant salon-pieces of the same kind. The piano figuration is of the type one finds again and again in Hummel and Weber, the trio being particularly Weberian.

Study of the rather later Variations sur un air national allemand and the first three Rondeaux is more rewarding, though they reveal little of Chopin's true musical individuality. But that does begin to emerge. The finale of the 'German' variations (*tempo di valse*) and the beginning of the D flat episode of the C minor Rondeau already give out a little of the unmistakable Chopinesque ring. One has only to compare the latter:

Ex. 1

with a very similar passage in the *allegretto* of Hummel's Sonata in A flat:

Ex. 2

to see both how the young Chopin belonged to his period and how he breathed a new poetic spirit into familiar forms. The end of the first sixteen-bar period of the *tempo di valse* of the Variations brings a formula:

Ex. 3

familiar in the more mature Valses (cf. Op. 34, No. 1, and Op. 42) and the coda, with its cantabile motive sung out in the tenor, sounds like a faint premonition of several passages in the Scherzi, notably the peroration of the C sharp minor. One also notes the appearance of one of Chopin's favourite chords: the diminished seventh on the sharpened fourth[1] (cf. the G sharp minor Polonaise, bar 10; the A minor Mazurka dedicated to Emile Gaillard, bar 2; and later works *passim*). (As we shall soon find, the diminished seventh plays an extremely important part in Chopin's harmony.) And so with other things: widely spread chords here (Op. 1, bars 102–3 and 106–7), a sequence there (*ibid*., bars 132–5).

These are portents which we see to be portentous only because we know what they portended. But broadly speaking the C minor Rondeau and the Variations on a German Air—and, for that matter, the slightly later but even less interesting Rondeau for two pianos—show Chopin merely at his taking-off point. His thematic material, his harmony, even his piano-writing—the figuration still usually bounded by the limits of the octave-stretch—are exactly what one would expect of a talented young contemporary of Hummel and Moscheles. And, incidentally, a composer who began in those days by imitating Hummel and Moscheles had not chosen bad models. Hummel in particular, though not a great master, was something more than a mere talented imitator of his first teacher, Mozart. It is true his music is often platitudinous and usually rather square-cut. But it has exquisite polish. (Hummel studied not only under Mozart, but

[1] In its second inversion, i.e. what old-fashioned English theorists would consider a third inversion of the supertonic minor ninth. Here, as so often later, Chopin employs convenient rather than correct notation: F double sharp instead of G natural.

under Clementi, Albrechtsberger, and Salieri, so it is hardly surprising that he was a master of the technique of composition.) If his figuration is often conventional—as Beethoven's piano-writing often is, too, for that matter—it sometimes assumes forms that suggest the mature Chopin himself, as in these two cases in the Sonata in F sharp minor, Op. 81:

Ex. 4

and

And in such passages as these from the B minor Concerto, Op. 89:

Ex. 5

and

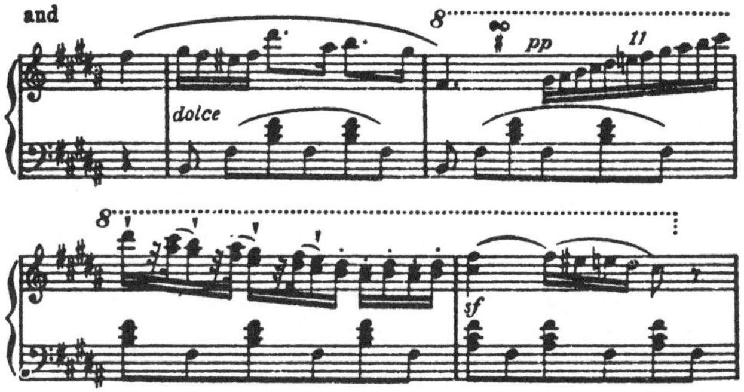

his idyllic romanticism is hardly distinguishable from Field's, and he comes fairly near to the mature Chopin of such Nocturnes as Op. 9, No. 3, and Op. 32, No. 2. He even uses the marking *tempo rubato* in the first movement of his C major Piano Sonata. He is in fact the chief link between Mozart and Chopin.[1]

Hummel's influence is much less apparent in the Rondeau à la mazurka, Op. 5, though, as we shall see, Chopin remained under it at least until 1830, when he wrote the E minor Concerto. But this Opus 5 is much more genuinely Chopinesque than anything that had preceded it. 'Who could fail to recognize him in the peculiar sweet and persuasive flows of sound, and the serpent-like winding of the melodic outline, the widespread chords, the chromatic progressions, the dissolving of the harmonies and the linking of their constituent parts!' cries Niecks. But that which especially

[1] More important even than Field, who was himself indebted to Hummel (cf. Field's Piano Concerto in A flat).

stamps the Rondeau à la mazurka with the mark of its authorship is—the fact that it is a mazurka. Here for the first time Chopin shows his intense love of the idiom of Polish folk music. It is Polish fresh air that blows away the Hummelian atmosphere of drawing-room and concert-platform. Weber, and Hummel himself, had written pieces embodying the rhythms of Polish dances—polonaises and so on—but none the less wholly salon-pieces. The Rondeau à la mazurka is a salon-piece too, but it is much nearer in spirit to the simple peasant dance. (Almost as near as Schubert in *his* dance music; indeed, despite Niecks's remarks, a good deal of the piano-writing in this piece is a good deal more like Schubert than true Chopin.) The very first bar shows that we are not to be fobbed off with superficial imitation of the mere rhythms of the mazurka. The key is F major, the third note a B natural.

Now the Lydian mode—the normal major scale with the fourth degree sharpened—is very common in Polish folk music.[1] Of the twelve examples of Polish folk music quoted by Niecks (admittedly 'chosen with a view rather to illustrate Chopin's indebtedness to Polish folk music than Polish folk music itself'), six have the sharpened fourths. The Mazurka in A minor, Op. 68, No. 2, written at about the same time as the Rondeau à la mazurka,

[1] And also found in Russian folk music, perhaps as the result of Polish influence. It is noteworthy, though perhaps only accidental, that Mussorgsky introduces the sharpened fourth in the polonaise of *Boris Godunov*.

also has the sharpened fourth, and we shall find it again and again in Chopin's later mazurkas. These harsh B naturals in the chief subject of the Rondeau are alone sufficient to sweep away any suggestion of that pleasant, superficially polished type of art suggested at the period by the name 'rondo'. (Even a rondo in 3/4 time, instead of 2/4, was something of a novelty.) But the implications of these B naturals go deeper than this: they show how thoroughly the young Chopin's musical mind must have been impregnated with the idiom of his native folk music if he could think in it as naturally and spontaneously as he does here. And there is the harmonic aspect too. After four bars of bare octaves Chopin begins to harmonize his melody. How then does he solve the problem of this awkward sharpened fourth? By harmonizing it with the very same characteristic chord that we have already noticed in the coda of the Variations sur un air national allemand: the second inversion of the diminished seventh on the sharpened fourth.[1] Is there not something more than coincidence in this hand-in-glove fitting of a favourite harmonic effect to a specifically Polish melodic effect? The chord itself was by no means new—Beethoven uses it in his very first piano sonata—but it is surely not too fanciful to suppose that Chopin's remarkable

1 Henceforth I do not propose to speak of inversions of diminished sevenths, or even of other chords, unless there is a special reason for doing so. A diminished seventh remains the same no matter how it is inverted, and inversions seldom affect the tonal functions—which are what matter principally—of other chords.

affection for it was due in the first place to its aptness to that modal peculiarity of Polish folk melody.

For the rest the harmony of the Rondeau is not yet strikingly individual. Yet there are more portents: for instance, the little chain of dominant sevenths in descending fifths, on G, on C and on F, leading back into the B flat theme, in bars 19–21 of the *tranquillamente e cantabile* section. As we shall see, such chains of dominant sevenths, each resolving, not on the expected simple triad, but on the expected triad *plus* an additional seventh, became a common feature of Chopin's mature harmonic style. Characteristic, too, is the *dolente* delay before the actual return of the graceful B flat tune; both the delay itself and the chord by which it is effected, an inversion of the minor form of the added sixth, are not only true romantic traits, but typically Chopinesque.

There are similar slight but significant portents in the actual piano-writing. Nothing yet very individual, but a tendency to more widely spread chords, a general—though only slight—loosening of the texture as if the 'prentice composer were getting more and more used to the feel of the sound-material under his hands, able to work with it more and more freely. But in this respect the Rondeau à la mazurka is far surpassed by the 'Là ci darem' Variations.

There is only the slightest element of virtuosity in the Rondeau, but the Variations sparkle with it from beginning to end. Let us be quite clear in our minds about this 'virtuosity'. Even in this avowed show-piece, Chopin's virtuosity is by no means that of the mere showman. His outward brilliance—and Liszt's, too, in a great many cases—is the natural manifestation of an inward exuberance, an exuberance generated by the consciousness of his own powers as a pianist. And if Liszt sometimes turns exuberantly to his audience, Chopin seems rather to be naïvely exulting in his command of the keyboard for his own private satisfaction. This inward spirit of virtuosity, this joy in freely and easily playing with one's sound-material and one's instrument, is manifest not only in the obvious bravura—the fioriture of the introduction, the triplet trills *staccato, ma leggiero e sempre più piano*, the tumultuous demisemiquaver rush of the second variation (anticipating the finale of the B flat minor Sonata), the bold, leaping *moto perpetuo* fourth, the dazzling finale—but in less spectacular traits: delicate washes of colour (bar 19 of the introduction and again the *delicato* passages just before the end of the introduction), the triplet counterpoint of the first variation, and above all the wide spread of the chords and of figures based on them.[1] In the introduction one finds chords spread over nearly two octaves in the left hand and, simultaneously, more than two in the right. From *Là ci darem* onwards, Chopin's figuration is no

[1] The second part of Variation 5, *cantabile e molto legato*, which so remarkably anticipates the love-scene in Strauss's *Heldenleben*, provides a curious instance of what one might call virtuosity of emotional expression. This is Don Giovanni, the supreme virtuoso in the technique of love-making.

longer octave-bound. That in itself necessitated a far more advanced pedal technique than anyone had employed hitherto. Chopin's own pedal markings for these Variations, given in the Oxford Edition, are of the greatest interest.

The variation form, of course, offered Chopin an easy way of escape from a problem that he never satisfactorily solved in these early days, and only sometimes and with difficulty in later years, except perhaps in a few works of his very latest period, a problem that he usually contrived to avoid: that of large-scale musical architecture. His youthful fondness for rondo form is symptomatic, for the rondo is potentially the loosest and most rudimentary of the larger musical forms—and Chopin makes the most of its possibilities. By comparison with the mature Beethoven sonata-rondo type, these early rondos of Chopin's, and even the finali of the Concertos, seem hardly musical organisms at all. They are essentially nothing but stringings together of smaller independent entities, though in the case of Op. 5, the all-pervading mazurka rhythm does give the piece a pleasant semblance of unity. We shall find Chopin later devoting special attention to the art of transition, which can skilfully conceal the joins of such sectional compositions. But even in the Second Impromptu, written ten years after the pieces we are discussing now, we shall see him failing lamentably, and in these early rondos the seams show all too plainly.

Of course, formal 'stringing together' instead of

organic form is inherent in the style of a composer who thinks lyrically, as Chopin mainly did. It is an extension on the grand scale of that process of balancing melodic phrase by melodic phrase which, with a lyrical composer, takes the place of the motive-evolution of a Beethoven. Admittedly, Chopin, even in his early works (e.g. the three Mazurkas, Op. 68, and the B flat subject of the Rondeau à la mazurka) shows a paradoxical predilection for melodic lines arising from the play of tiny motives instead of developing in free cantilena. But that, as Ernst Bücken was I think the first to point out,[1] is also a pecularity of the folk mazurka.[2] Chopin absorbed it into his musical system in the composition of his own mazurkas, and it quickly became one of the most striking traits of his melodic style. (The Grande Valse brillante, Op. 18, is an obvious example that one thinks of immediately, but Chopin's music is full of far more subtle and interesting cases of motive-generated melody, e.g. the superb Étude in E Op. 10, No. 3,

[1] *Die Musik des 19. Jahrhunderts bis zur Moderne*. Potsdam. 1929.

[2] For instance, the tune, said to be 'popular in the neighbourhood of Warsaw', illustrating the article 'Mazurka' in *Grove:*

Ex. 6

and the third-period Nocturne, Op. 55, No. 2.) All the same, a melody evolved from a single motive is still a melody, not a theme; the music of which it forms the backbone still remains lyrical so long as the composer closes his ears to its latent possibilities for development. And the young Chopin did close his ears.

The first movement of the C minor Sonata suggests that he may have convinced himself that he was incapable of development on Beethovenian lines; this very experiment proves that in these early days he was. He worked here with true themes, not with motive-inspired melodies, but even allowing for the fact that the material is incredibly dull and sterile—the whole Sonata is so evidently a student-exercise that it is difficult to understand why Chopin should have sent it to a publisher[1]—his complete inability to develop anything from it is really remarkable. Ernest Newman's celebrated jeer at 'the average academy-made composer with his tiresome and futile attempts to make living music by the mechanical manipulation of a couple of arid "subjects", reminding us of nothing so much as some poor patient Hottentot rubbing two dry sticks together in the hope of getting a bit of fire', might well have been suggested by Chopin's first movement, except that Chopin rather suggests an insane Hottentot rubbing a stick against nothing at all.

[1] Who, however, had the good sense to forget it for twenty-three years, and published it only in July 1851, two years after the composer's death.

The whole Sonata is lifeless, and—with the exception of a few bars in the second section of the trio that suggest the Chopin of the valses and mazurkas—completely un-Chopinesque. Even the piano-writing is extraordinarily dull and conservative; perhaps because the young composer felt that any suggestion of virtuosity was incompatible with pseudo-classical sonata-composition.

The whole work demonstrates both that the young Chopin's mind was incapable of generating living musical substance from a compact theme, and that he had no conception, other than the driest textbook conception, of the first principles of sonata-form. If any confirmation were needed—it is not—one would find it in the first movement of the G minor Trio, which we know was written in 1828; from the textbook point of view the student had made considerable progress—and he certainly wrote better music here—but he was just as far as ever from organic development and organic form.

The two other big show-pieces for piano and orchestra dating from this period, the Fantaisie sur des airs nationaux polonais, Op. 13, and the Krakowiak (Grand Rondeau de Concert, Op. 14) both dodge this problem of form in the ways indicated by their titles. The rondo was to the young Chopin an ever present help in time of structural trouble. As for the Fantaisie, it does not make even a decent pretence of being anything but a string of variations on one theme after another, each section being cobbled on to the next with hardly a semblance of art or artifice, and each being happily forgotten once it is done with. And yet these two show-pieces, written simply as vehicles for the composer's virtuosity, slight as their artistic value may be, are by no means without interest. They are deservedly forgotten by pianists but they must not be overlooked by the student of Chopin's slowly maturing style.

They are full of the lights and shadows of coming events. Not of actual passages of course, though, as Niecks points out, the first variation on the air 'Juz miesiac zaszedl' ('The moon had set') in the Fantaisie 'contains the germ of the charming Berceuse'; but full of anticipations of characteristic technical devices in the harmony and the *facture* for piano. The bravura writing is free and dashing but, except for the now almost invariable wide spread of the chords and figures, not remarkable. But the filigree ornamentation of the quieter passages already suggests the Chopin of the later nocturnes much more definitely than anything in the only slightly earlier Nocturne in E minor, Op. 72, No. 1.

The *largo* introduction to the Fantaisie particularly repays study from this point of view. The essence of Chopin's filigree style, as of Field's and Hummel's, is the throwing of a lacy and mainly chromatic veil by the right hand over a firm basis of diatonic broken chords in the left. Though Chopin's left-hand chords are more widely spread and hence more sonorous than Hummel's (but not more than Field's frequently are), so long as he is content to draw filmy single-note arabesques in the treble, *delicatissimo* and *leggierissimo*, as he does in this introduction, he is simply copying Field and Hummel. Even his irregular groupings of demisemiquavers in thirteens and thirty-threes over steadily rocking quavers were no innovation. In the *Larghetto a capriccio* of his Sonata in D major, Op. 106, Hummel had spun chromatic fioriture in groups of 37, 34, 17, 19, and 15 demisemiquavers that are more Chopinesque than a good deal of early Chopin. (On the light Viennese pianos that Hummel played, such effects were easy and exquisite; Chopin, too, played on and wrote for pianos with a much lighter touch than ours—on which his silvery washes of delicate colour suffer transmutation, without the greatest care, into vulgar display passages.) But in the introduction to the Polish Fantasia Chopin no longer limits himself to simple fioriture. He had already hinted in the introduction to the 'Là ci darem' Variations (bar 19) that he knew other ways of chromatically embroidering a plain diatonic chord; now he freely employs double notes and introduces a new and altogether delicious appoggiatura technique:

Ex. 7

One passage near the beginning is specially prophetic of the mature Chopin:

Ex. 8

The emphasis obtained by the combination of the quintuplet and the novel flowering into significance of the inside part is as characteristic as the unexpected sharpening of the E and the consequent enharmonic twist on to the subdominant minor chord.

In both these examples the music is as fundamentally diatonic as anything in Hummel. But in one or two passages of the Fantaisie and Krakowiak —and at the peroration of the 'Là ci darem' Variations, sixteen bars before the end—Chopin breaks up the diatonic surface into a coruscating shower of chromatic particles:

Ex. 9

(a) Là ci darem

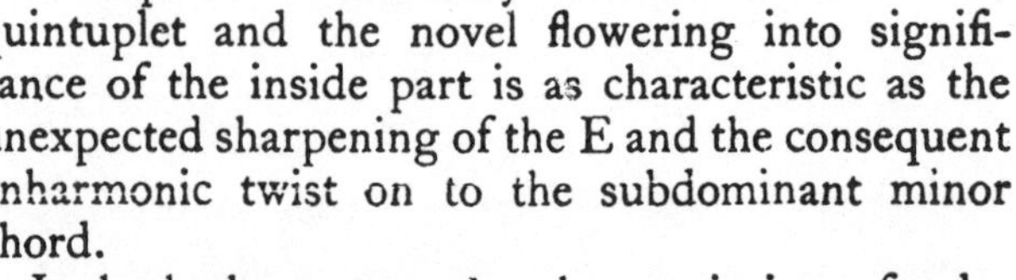

(*b*) **Polish Fantasia**

(*c*) **Krakowiak**

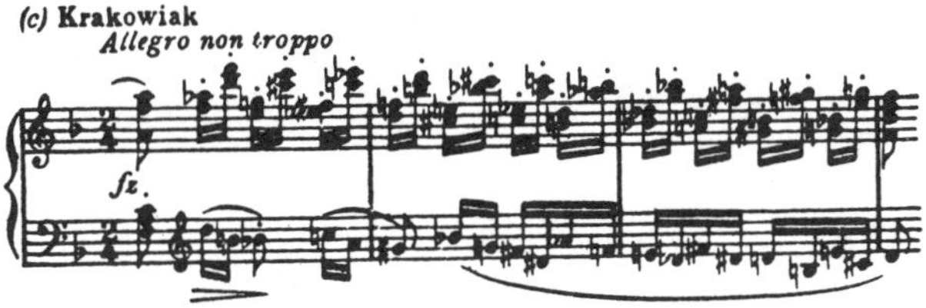

In each case the same thing has happened. To change the metaphor: Chopin has stepped from firm diatonic ground on to his favourite diminished seventh on the sharpened fourth, and taking off from that slippery point, danced giddily in the air through a sequence of diminished sevenths from the last of which he springs adroitly back to firm ground. In two of the three cases the chord is simply 'side-slipped' bodily, semitone by semitone; in Ex. 9*b* the effect is subtilized by repeating each chord in an inversion and then making the semitonal side-slip. In each case the effect is the same. It is notorious that any diminished seventh is tonally ambiguous; its key can be determined only from its context. And here we have a whole series of diminished sevenths of which only the first and last can be even fleetingly related to a key at all. In other words there has been a temporary suspension of the principle of tonality. Atonality, at any rate as a passing phenomenon, has become a fact.

Indeed it had done so already long before Chopin. As long ago as 1763 Haydn had ended the exposition of a Symphony in E (No. 12 in the Breitkopf *Gesamtausgabe*) with a firm dominant unison B and opened the development with this mysterious passage:

Ex. 10

Three years later we find Jommelli writing in his *Vologeso*:

Ex. 11

and again in Gluck's *Iphigénie en Aulide* (Clytemnestra's 'Jupiter, lance ta foudre') (1774) and in Méhul's *Ariodant* (1799):

Ex. 12

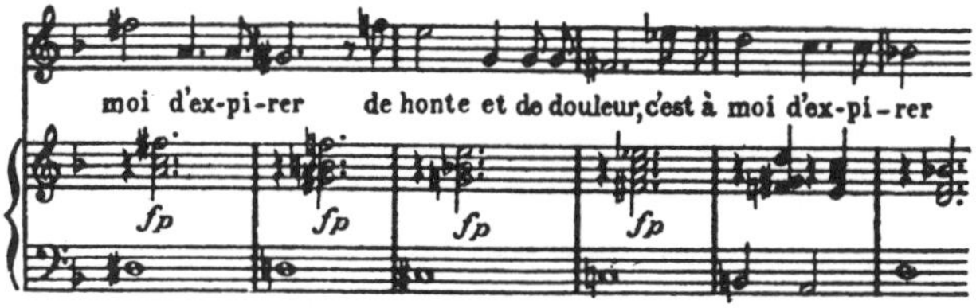

The almost consonant sound of the chord made such successions easily acceptable to the ear, and even Chopin's admired Hummel, who was no harmonic revolutionist, has

Ex. 13

in his early Sonata in E flat, Op. 13.

But few things are more fatuous than a discussion of technical processes that leaves out of account the spirit in which the processes are applied. This particular device was not new in itself, but Chopin's use of it was quite unprecedented. Haydn had employed it as a 'surprise' effect; it must be one of the earliest of his numerous attempts to *épater les aristocrates.* Gluck and Jommelli and Méhul use it to emphasize points of dramatic intensity. With Hummel it is (shall we say?) purely musical, a touch of colour only in the sense that it does relieve the penny-plain E-flatness of its surroundings with a passing touch of twopenny chromaticism. Chopin, too, uses it in a purely musical sense. But how differently! In his pianist's hands it becomes a prism breaking up the white diatonic light into glittering rainbow particles. It has now become purely pianistic, reborn under a virtuoso's fingers and possible only in terms of piano timbre. And if still only a parenthetic effect, it is no longer so briefly transitory as in Hummel; in the Variations it lasts for four bars, though only two are quoted in Ex. 9*a*. The tiny crack in the firm tonal surface has widened perceptibly.

But filigree ornamentation and interesting little splashes of harmonic colour do not make masterpieces. The Fantaisie and the Krakowiak are simply bravura show-pieces, inferior in musical value even to the 'Là ci darem' Variations. If the introduction to the Fantaisie contains a number of the technical ingredients of the later nocturnes, it is almost completely devoid of their poetry. One must agree with Niecks that 'the best parts of the works, those that contain the greatest amount of music, are certainly the exceedingly spirited Kujawiak' (the last section of the Fantaisie) 'and the Krakowiak'. In other words, those based on folk material. For the Kujawiak is, on the admission of the title, an *air national polonais* and the principal theme of the Krakowiak is a variant of one quoted by Grove[1] as the tune to which the celebrated Fanny Elssler danced the *cracovienne* about 1840. 'Whether it is a real krakoviak, or a mere imitation, the writer is unable to say', he remarks. It is possibly a debasement, and Chopin's version a refinement, of a common original. On the other hand, both may be true folk variants of the kind familiar to students of the folk music of all countries.

The kujawiak is a species of mazurka—according to all the authorities the slowest kind of mazurka, though Chopin marks his *vivace*—and the non-bravura parts of this specimen are so like Chopin's early mazurkas that one sees immediately, even if one has not realized before, how close he remained in these works to authentic folk melody. And not to melody alone. The influence of the drone bass, also characteristic of the folk mazurka, is apparent, and the influence of the Lydian mode peeps out in the harmony (bar 7) long before it appears for a moment in the melody (bars 29 and 31). But before long the naïve peasant dance loses its character and becomes a brilliant concert-piece; the pianist has been too much for the Pole.

The mazurkas actually so-called written at this time, the three published posthumously as Op. 68, Nos. 1, 2, and 3, are completely devoid of any element of virtuosity. They are piano music only in the sense that Schubert's Ländler are piano music and might easily be simple transcriptions of

[1] *Dictionary of Music and Musicians* (3rd ed.) Article, 'Krakoviak'.

authentic peasant mazurkas. Between them they represent the three main types of mazurka (all characterized by the accent on the third beat[1]): the quick oberek or obertas (No. 1), the slow and melancholy kujawiak (No. 2), strikingly unlike the example so-called in the Fantaisie, and the intermediate type that has given its name to the whole family, the mazur (No. 3). All three possess the characteristics of the folk mazurka: love of the sharpened fourth (opening of No. 2 and middle section of No. 3), introduction of triplets in the melody (No. 1), play with tiny motives (all three), drone bass (whether stylized into an unobtrusive pedal as in the opening of No. 2 or emphasized in primitive open fifths as in the middle sections of the same piece and of No. 3), the feminine ending of the piece (even if only suggested by the left hand as in No. 2), and of course the characteristic rhythms and melodic patterns of the folk mazurkas throughout.

Nothing could be squarer than the construction, entirely in four- or eight-bar periods. The harmony is as impersonal as the piano-writing. Only a tendency to thicken out melodies in sweet thirds and sixths—also noticeable here and there in other early works—must have come from the duets of contemporary Italian opera rather than from the music of the Polish peasants. Yet, with all this squareness and a lack of finish amounting almost to crudeness, despite the complete absence of

[1] According to some authorities, the mazur takes the accent on the second beat.

Chopinesque piano-poetry, one feels that this is the real Chopin, working not only in a congenial medium but on a scale that in no way embarrasses him. These early mazurkas have none of the polish of the good miniature, yet form and content perfectly match—if only because the form simply arises (as it should) from the naïve, square-cut content.

The three polonaises, Op. 71, written at the same time as these mazurkas (1827–29) and, like them, published only posthumously, stand in the same relation to the later polonaises as their companions to the mature mazurkas. And the polonaises in general are more interesting in structure, harmony, and pianistic texture than the mazurkas. Whereas the mazurka is a naïve peasant dance, the polonaise was sophisticated, the dance of the Polish aristocracy—indeed less a dance than a processional ceremony—and this is reflected in Chopin's music. The polonaise offered no folk idiom to be absorbed, but all the same it observed certain well-marked conventions which Chopin adopted either simply or in stylized forms. Apart from the all-pervading accompaniment rhythm  there were certain rhythmic formulae for the end of the melody: e.g. 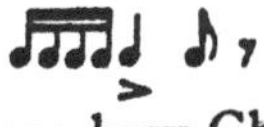 or  and it is interesting to see how Chopin has refined and subtilized the latter figure in the final bars of all three of these polonaises. Another characteristic of the traditional polonaise is a tendency to repeat motives:

Ex. 14

a process not unrelated to the slightly different motive-play of the folk mazurka. This, too, is reflected in Chopin, and more clearly in these early polonaises (e.g. the opening of Op. 71, No. 2) than in the more mature ones. On the other hand, the influence of Weber's polonaises is not yet entirely thrown off, though it is much less obvious than in the G sharp minor Polonaise of 1822; even the figuration suggests a mingling of Weberian with purely Chopinesque traits.

One notes the shimmer of chromatically descending diminished sevenths again (near the end of the D minor) and the reappearance in a more subtle form (in Op. 71, No. 2) of the chain-of-dominant-sevenths device already noticed in the Rondeau à la mazurka[1] (bars 20–6 of the middle section). And the modulation made by taking the last dominant seventh of the chain as a German sixth in the new key became one of Chopin's favourite bits of sleight-of-hand. Eight bars after this last-mentioned passage (i.e. bars 18–19 before the *da capo*) occurs another typically Chopinesque harmonic effect: the dominant seventh changes from first inversion to root position by semitonal movement of the outside parts, thus producing transitional chords that have no tonal function.

[1] See p. 10.

Structurally these polonaises are very simple but less primitive than the mazurkas of the same period. While the latter are cast in the simplest ternary form, ABA, modified only by a slight curtailment of the repeat of the A section in Op. 68, Nos. 2 and 3, and keep within a very limited range of keys, the polonaises show that form carried to a higher power, as it were, each section being itself in ternary form. The range of keys in each piece is much wider and the second part of the central section of the B flat Polonaise—the crown of its structural arch and the part containing the interesting harmonic features just mentioned—is a sort of rudimentary development section. As in the mazurkas the dance-form 'excuses' the abruptness of the transitions, while the all-pervading dance rhythm unifies the whole.

Neither the three Valses (Op. 69, No. 2, Op. 70, No. 3, and the posthumous E minor) nor the single Nocturne (Op. 72, No. 1) of this period possess any interest beyond that attaching to early essays in forms in which Chopin was later to create masterpieces. Chopin's valses as a whole are much simpler harmonically than his other compositions. But even here an important part is played by harmonic effects for their own sake. Consider the second section of the E minor Valse; it begins with two apparently quite different eight-bar phrases which prove on examination to be both based on the same harmonic progression. The original conception must have been harmonic, the quasi-melodic figuration an afterthought. Both the motive-generated melodic pattern of the first phrase and the sequential treatment of the motive are characteristic of Chopin's melodic style. (Compare also bars 13–15 of the same Valse and the middle section of Op. 70, No. 3.)

Formally, too, these three Valses are extremely simple, though the E minor—in every way the most interesting—shows Chopin in more open revolt against the rule of symmetry in ternary form. In the Mazurka, Op. 68, No. 3, trusting to that principle of musical form which has been aptly compared with perspective foreshortening, he had cut down the repeat of the A section by half; sixteen bars heard now do quite satisfactorily balance thirty-two bars heard earlier, at a little distance in the memory. In the E minor Valse Chopin carries this principle still further by drastically reducing the repeat of A to a mere eight bars—extended, it is true, by a twenty-seven-bar coda arising out of them. The Nocturne in the same key is also extremely simple in build, though here it is the middle section that is disproportionately short[1] and the repeat is given mild additional interest by the not very characteristic ornamentation of the melody. The nine-bar phrase that concludes the first section at least shows that Chopin was aware of the booby-trap of square phrase-building in these small lyrical forms.

Otherwise, this first of all the nocturnes is curiously uncharacteristic. In style—*cantabile* right-hand melody over quiet, wide-spread triplet

[1] In some of the later Nocturnes (e.g. Op. 9, No. 1, and Op. 27, No. 1) the middle section is disproportionately long.

arpeggios—it is hardly distinguishable from a number of Field's nocturnes (e.g. Nos. 1, 2, and 5) and it lacks Field's peculiar idyllic charm. It appears to be a composition exercise. For the earliest specimens of the true romantic Chopin nocturne, we must turn to the slow movements of the two Concertos.

With the Concertos we at last reach the earliest of Chopin's works that have kept their places in the ordinary repertoire. Not very prominent places, perhaps; they are immature and even the warmest of Chopin's admirers must admit that they are poor examples of the piano concerto as such. But they—particularly their slow movements—contain far too much beautiful music to be allowed to follow the 'Là ci darem' Variations and the Fantasia on Polish Airs into the oblivion of dictionaries and history books.

It is useless to look in these two works for the beauties of concerto-form that one finds in Mozart and Beethoven. Like the Variations and Fantaisie and Krakowiak, they must be considered purely and simply as large-scale show-pieces for the soloist, with necessary but regrettable orchestral backgrounds. After the opening ritornello the orchestra's share is more or less limited to brief interjections and the provision of a light harmonic support to the solo. Niecks said of the orchestral accompaniments of the three earlier pieces that they show 'an inaptitude in writing for any other instrument than the piano that is quite surprising considering the great musical endowments of

Chopin in other respects'. And that is true also of the Concertos—where the inevitable long tuttis show up the inadequacy of the orchestration far more pitilessly than any of the short passages in the earlier works—and of the Grande Polonaise brillante for piano and orchestra written at about the same time. It is significant that after this Chopin never wrote for orchestra again.[1]

The perfunctory treatment of the orchestra in the Concertos, however, must be attributed not only to Chopin's unhappiness in handling it, but to the example of his model, Hummel. Hummel also had given the orchestra a merely subordinate role in his concertos and his lead was followed not only by Chopin, but by most of the second- and third-rate concerto composers of the romantic dawn: Field, Kalkbrenner, Moscheles, and the rest. Fundamentally, this school derived from Johann Christian Bach. In the hands of the great Bach's sons, the genre *Klavierkonzert*[2] had split into two main types. C. P. E. Bach, who kept faithfully to

[1] One peculiarity of Chopin's scoring is worth notice: the puzzling inclusion in both concertos of a part for a single trombone. It can hardly be a mere coincidence that Glinka's *Night in Madrid* and *Kamarinskaya*, also written specially for a Warsaw orchestra, though nearly twenty years later, likewise contain parts for a single bass trombone. Obviously Warsaw possessed only one respectable trombonist. But he must have been a good player; Glinka tells us in his memoirs that the Governor's *Kapellmeister*, Pohlenz, 'orchestrated my *Prayer*' (a piano piece) 'with trombone obbligato and it was not ineffective'.

[2] Which had come into existence in Johann Sebastian's generation as an offspring of the violin concerto.

the true concerto idea, the interplay of solo and tutti each equally important, stands at the head of the great line of heredity that runs through the later Mozart and Beethoven to Brahms. On the other hand, Johann Christian Bach, who sacrificed the dialogue principle to the importance of the solo instrument, which he provided with brilliant passage-work as well as a half-Italian cantabile type of melody (and who incidentally introduced the rondo finale), became the forefather of the virtuoso concerto of the younger Mozart, Hummel, Field, and Chopin. And of this type in the early nineteenth century Hummel was indisputably the most skilful practitioner.

Arnold Schering[1] suggests a non-Hummelian model for Chopin's E minor Concerto: the Grand Concerto in D minor, Op. 61, of Kalkbrenner to whom the Chopin work is actually dedicated. He draws attention to 'a certain relationship between Kalkbrenner's opening:

Ex. 15

and Chopin's, particularly noticeable in the bass', and goes on to point out other features common to both—but, unfortunately for his theory, also to a good many other concertos of the same period. If Schering had not been led astray by the dedication of the E minor Concerto, he might have

[1] *Geschichte des Instrumentalkonzerts.* Second Edition. Leipzig, 1927.

noticed that the opening of the earlier F minor work,[1] with its minor thirds descending chromatically, is really more akin to the Kalkbrenner. And if he had looked a little more closely at the opening of Hummel's Concerto in A minor, Op. 85, he would have found an actual identity of certain motives in it and in the opening of the Chopin E minor:

Ex. 16

In this light Chopin's theme is seen to be essentially an elaboration and extension of the older master's and at the same time a considerable improvement on it, longer-breathed, rising and falling in one big eight-bar curve. Notwithstanding which, Chopin was completely unable to evolve from it a living musical tissue.

Chopin's technical kinship with Hummel[2] and

[1] The later opus-number of the F minor and its general superiority are misleading. Op. 21 was composed in 1829; Op. 11 in 1830.

[2] He even writes for strings *col legno* in the finale of the F minor as Hummel had done.

his immense superiority are even more clearly demonstrated by comparison of the *larghetti* of both concertos with the corresponding movement of Hummel's B minor Concerto. Technically all three can be described in the same general terms: a brief orchestral preamble serves to introduce a prolonged solo of the nocturne-type, based on a simple melody (quickly flowering into delicious fioriture) played by the right hand in single notes or octaves over a rocking quaver accompaniment in the left.[1] But Hummel's music is prose, Chopin's is poetry—sentimental, if you like, and inferior to his later poetry, but undeniably poetry. Hummel's is classical, or rather an imitation of the smooth outward form of classicism, Chopin's is romantic. Or in the more precise terms of style: Hummel's melody is square-cut and centres feebly on the

[1] Structurally both the Chopin slow movements are timid extensions of the ABA formula:

F minor:

Brief orchestral prelude.
A—Main theme (solo) in A flat.
A—The same repeated with further decorative elaboration (solo) (A flat).
B—Recitative for piano in octaves (A flat minor).
A—Main theme with fresh ornamentation (solo) (A flat).
Brief orchestral postlude (on same material as prelude).

E minor:

Brief orchestral prelude.
A—Main theme (solo) in E major.
B—Secondary theme in B major (solo).
A—Main theme ornamented in E major (solo) and modulating through brief interludes in G sharp minor to—
B—Secondary theme in G sharp major (solo).
A—Main theme in orchestra, accompanied by piano figuration (E major).

notes of the common chord,[1] Chopin's—at least in the F minor—pours out in a long-breathed, sensitive, passionate cantilena proceeding mainly in conjunct motion like a vocal melody and ornamented by exquisite piano-coloratura. We are face to face, for the first time, with the miracle of Chopin's nocturne-type of melody which *is* vocal in essence, originating in a very different type of music from Hummel's: Italian opera in general and Bellini's in particular. Chopin's general indebtedness to Bellini will be discussed later but it is worth pointing out here in passing that the operatic flavour of the F minor slow movement is heightened by its recitative-like middle section, the accompaniment to which (tremolo string chords with ominous double-bass pizzicati) is, as Tovey says,[2] 'as fine a piece of instrumention as Berlioz could have chosen to quote in his famous treatise'.

The final movements of both concertos are likewise Hummelian rondos: the F minor by Hummel out of the mazurka, the E minor even more Hummelian and rather superficial in its charm, though with one touch of humour almost worthy of Beethoven—the reappearance of the principal theme a semitone flat, a 'mistake' which is not corrected till eight bars later. Chopin repeated the jest some years later in the F sharp Impromptu.

But in the concerto, as in the symphony, it is the first movement by which a work stands or falls. And in the structure of their first movements the

[1] As does that of the *larghetto* of his A minor Concerto.
[2] *Essays in Musical Analysis.* Vol. iii. O.U.P. 1936.

two Chopin concertos, which neither stand nor fall but stumble, do not follow Hummel as slavishly as seems to be generally supposed. In both his A minor and his B minor Concertos Hummel works to precisely the same pattern, a pattern extraordinarily like that of Beethoven's much greater C minor. In that fictitious abstraction which the authors of textbooks on musical form call 'classical concerto form', the opening tutti presents both subjects[1] in the tonic key (as indeed is usually the case in Mozart), leaving them to be sorted out and tonally distributed by the soloist; Beethoven in the C minor and Hummel in both these concertos put their second subjects at once in the relative major in the first tutti. They agree also in winding up the first tutti with an extensive return to first-subject material in the tonic. Chopin follows them in the F minor in at once stating his second subject in the relative major, but not in returning to the first-subject material. In the E minor he makes the return to first-subject material but puts the second subject in the *tonic* major. In both he finally recapitulates the second subject in the relative major: in other words, in the F minor the reprise tamely follows the same unenterprising key-sequence as the exposition—the flouting of a basic principle of sonata-form—while in the E minor the same principle is turned completely on its head, the twofold exposition clinging stoutly to the tonic throughout, with a perverse shifting of tonic in the

[1] By 'subjects', of course, I mean not 'themes', but whole sections of music, each of which may contain several themes.

recapitulation. And whereas a greater master of form might have contrived to commit all these crimes and disguise them as strokes of genius. Chopin merely demonstrates that the beauties of key-relationship and key-balance perceived by all the Viennese classical masters meant nothing to him. Let us be thankful for that; if they had meant a great deal to him, he might have given us two rather better concertos but he would have found it immensely more difficult later to break up that solid diatonic system into the fascinating, constantly modulating, frequently chromatic sound-texture which constitutes about three-quarters of the real Chopin and which opened the way for Liszt, Wagner, Debussy, and Schönberg.

Chopin's weakness of key-sense was equalled only by the weakness of his sense of development, the other main feature of the great classical concertos. The nearer he approaches to a passable imitation of true development, the drier and less Chopinesque the music becomes. Instead of growing up from the themes, it dwindles down from them. Nothing flowers from them, even from the promising opening of the E minor; Chopin can only strive desperately to squeeze something out of them. In the so-called development sections, falling back on Hummelian precedent, he abandons even the pretence of working out; the music consists simply of brilliant passage-work for the soloist against which, in the F minor, the orchestra bandies about fragments of the first-subject material mainly on solo wind. In the E minor thematic relevance is limited to the employment for about a dozen bars of the opening of Ex. 16*b* as a bass. As Niecks puts it, 'the pianoforte part reminds one too much of a study, without having the beauty of Chopin's compositions thus entitled'. The concertos as a whole, however, do show Chopin's keyboard writing—his filigree fioriture and his characteristic widespread passage-work—at the highest point it had reached so far. Technically they form together a complete encyclopaedia of the new keyboard idiom and a rich foretaste of the new harmonic idiom that developed in its train: dizzy chromatic side-slippings of dissonances, analogous to Ex. 9, the hard glitter of seventh-producing appoggiaturas, as at the end of the first movement of the E minor, and so on. But all this can be studied much more conveniently and to more purpose in the set of twelve Études, Op. 10, which are in so many respects quite mature and so fine that it seems a little unfair to class them with the works of Chopin's nonage merely on chronological grounds.

The Op. 10 Études enable one to see at a glance how far Chopin's technique had advanced beyond his predecessors by comparing each study with a slightly older one dealing with the same technical difficulty.[1] The very first in C major, for instance,

[1] The dedication, 'à son ami, F. Liszt', is not a tacit admission of indebtedness, by the way. Liszt's earliest studies, other than the boyish *Études pour le piano en douze Exercices* of 1826, did not appear till five or six years after Chopin's Op. 10, later even than his Op. 25.

with its tremendous striding arpeggios, covers the same technical ground as Cramer's Study in D minor (No. 19 in von Bülow's edition) but where Cramer's groups lie within fifths and sixths, Chopin's fill octaves; where Cramer asks for an octave stretch, Chopin demands a tenth. Similarly the left-hand accompaniment arpeggio in No. 9 covers elevenths and thirteenths as compared with Cramer's tenths and sixths in the parallel figure of his D major Étude (No. 53). And Chopin's No. 7 and No. 10, both very characteristic of his piano-style, similarly complicate the problems of Cramer's No. 52:

Ex. 17

and Clementi's Étude No. 23 (Tausig's edition):

Ex. 18

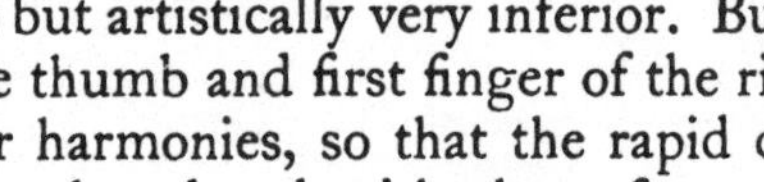

Again, more than one critic has drawn attention to the affinity between No. 2 in A minor and Moscheles' Étude in G, Op. 70, No. 3, likewise chromatic but artistically very inferior. But Chopin ties up the thumb and first finger of the right hand with inner harmonies, so that the rapid chromatic scales must be played with three fingers only: a reversion, as Leichtentritt points out, to a type of fingering used in the pre-Bach period. The enormous importance Chopin attached to this Étude is shown by the fingering, which he has given for every note except in repeated passages; no other composition in the whole of his music is so completely fingered; none demonstrates more clearly the importance of following the composer's own fingering and not that of his ingenious editors. The basic principle will be found to be the almost invariable use of the third finger (Continental style) for all black keys.

But the tackling of such awkward figures as those in Ex. 31, from the F major Étude, No. 8, and the tremendous spread chords in both hands of No. 11, were as completely new technically as were the profound tragic power of No. 12 and the sheer melodic loveliness of No. 3[1] musically.

The technical purpose of an étude can best be achieved by embodying the particular problem in a single motive or pattern which is persistently worked out all through the piece. And this basing of the whole composition on one motive, which has always been a peculiarity of the étude, solved for Chopin the real problem of miniature form: contrast and variety within a small unity. Even the purely melodic studies, Nos. 3 and 6, are based mainly on motive-generated melodies with regular-

[1] Speaking of No. 3 in E major, Niecks tells us that 'Chopin said to Gutmann that he had never in his life written another such beautiful melody'. Rimsky-Korsakov considered it the most perfect melody in all music, 'the very music of the Kingdom of Heaven'.

patterned accompaniments, and the all-pervading figure unifies each piece as do the dance-rhythms of the polonaise or mazurka, and does it much more organically. (Exactly as such Schubert songs as 'Auf dem Wasser zu singen' and 'Gretchen am Spinnrad' are held together by the accompaniment figure.) And whereas the clear-cut phrases of the dance and the alternations of its contrasting melodic ideas draw attention to joins and to the inevitable squareness of structure, the plastic nature of the single motive and the *moto perpetuo* character of most of the patterns in an étude conceal the joins and enormously simplify that business of transition which is one of the subtlest problems of the art of musical composition.

The broad outlines of these Etudes are simple enough. Like their successors in Op. 25, they are almost all cast in the simple, slightly modified ABA form, with or without coda, of the valses and mazurkas. But the middle section is usually marked not by new material but by contrast of key, colourful modulation, and varied treatment of the initial motive and technical problem; in fact, is not so much a trio as a miniature development section. There can be no development in the true sense, for there is usually no 'theme' to develop. But there is plenty of scope for that unconstrained, more or less improvisatory play with pianistic figures and colourful harmonic progressions in which Chopin delighted.

The great C minor, the so-called 'Revolution' Study, represents Chopin's supreme formal achievement up to the date (September, 1831) at which it was written. At least half a dozen different left-hand figures are employed, but they are all marked by regular semiquaver movement and this, together with the tragic declamatory right-hand melody, holds the whole together. The piece sounds as if dashed off in one great outburst of passionate feeling. The modulating middle section is swept in so naturally and leads so naturally back into the repeat, now wonderfully intensified; the repeat of the first section grows so naturally into the overpoweringly effective coda, the melody being simply augmented in double note-values; and this coda is so superbly conceived: that the simple underlying ternary form is quite unnoticed. Yet the piece is actually almost as squarely built as any of the early mazurkas:

A—8 bars introduction.
Phrase of 2+8 bars.
The same repeated, with the last 4 bars modified (say, 2+4+4) leading into—
B—Three 4-bar units, the last leading back into—
A—8 bars introduction.
Phrase of 2+8 bars.
Phrase of 2+4+4 bars, the last four being again different.
Coda—8 bars (growing out of the previous four by augmentation of the motive).
4 bars based on the introduction figure, *sotto voce*.

4 bars based on the introduction figure, *ff ed appassionato.*

The form is as simple as ever but it is now employed by a master instead of by a novice.

Harmonically, too, several of the Op. 10 Études announce Chopin's arrival at full mastery, so, instead of examining them separately from this point of view, I propose to group them with the other works of maturity in a general discussion of Chopin's harmonic style (pp. 77–95).

II

CHOPIN'S MATURE STYLE

(1831–1840)

THE compositions of Chopin's second period, those written after his arrival in Paris in 1831 and before the last change in style perceptible from about 1841 onwards, are as follows:

Ballades in G minor, Op. 23 (1831–5), and F, Op. 38 (1836–39)
Bolero, Op. 19 (1833)
Études, Op. 25, and Trois Nouvelles Études (1832–39)
Fantaisie-Impromptu, Op. 66 (1834)
Impromptus in A flat, Op. 29 (1837), and F sharp, Op. 36 (1839)
Mazurkas, Op. 17 (1832–3), Op. 24 (1834–5), Op. 67, Nos. 1 and 3 (1835), Opp. 30 and 33 (1836–38), Op. 41, (1838–9), in A minor (*Notre temps*) (1840).
Nocturnes, Op. 15, No. 3 (1833), Op. 27 (1834–5), Op. 32 (1836–7), Op. 37 (1838–9)
Polonaises, Op. 26 (1834–5) and Op. 40 (1838–9)
Preludes, Op. 28 (1836–39)
Rondo in E flat, Op. 16 (1832)
Scherzos in B minor, Op. 20 (1831–2), B flat minor, Op. 31 (1837), and C sharp minor, Op. 39 (1839)
Sonata in B flat minor, Op. 35 (1839)

Valses, Op. 18 (1831), Op. 69, No. 1 and Op. 70, No. 1 (1835), Op. 34 (1831–38), Op. 42 (1840)

Variations brillantes (on an air from Hérold's *Ludovic*), Op. 12 (1833)

Some of these are of little importance. The Bolero, the E flat Rondo, and the Op. 12 Variations can be entirely left out of account, though even the least characteristic work of the three, the Variations, contains a passage, the *lento* variation in D flat, that could hardly have been written by anyone else. It is bad Chopin; just how bad may be seen very easily by comparing it with the D flat Nocturne, Op. 27, No. 2; but unmistakable Chopin for all that. The remaining compositions of this period include much of his finest and most characteristic work. It is no longer necessary to trace his artistic development step by step. We can stand back a little and look at this output as a whole from various points of view: form, melody, harmony, and keyboard style in general.

Chopin's form is generally considered to be his weakest point. It was also the weakest point of all his contemporaries. And, of course, compared with Beethoven's—a fantastically unfair comparison—his sense of form is primitive, being limited almost exclusively to the possibilities of more or less modified ternary form. The elementary formula ABA is the structural basis of the vast majority of Chopin's shorter pieces. It underlies the majority of his second-period mazurkas, all his second-period polonaises,[1] and practically all the second-period nocturnes. Its predominance in the études has already been mentioned—Nos. 1 and 3 of the Trois Nouvelles Études are the only exceptions—and it is either implicit or explicit in a number of the preludes. Only the valses tend to break away into looser patterns akin to the 'suite of waltzes' initiated by Hummel in 1808 (*Tänze f.d. Apollo Saale*, Op. 31) and familiar through the works of the Strausses, Lanner, and Gung'l. Ternary form is, as we shall see presently, even the basis of the larger forms evolved by Chopin: the scherzo, the impromptu, and the ballade.

But, admitting the primitive nature of Chopin's basic conception of form, one can have nothing but praise for the skill with which he so often modifies, adapts, or even completely conceals this naïve basis. We have already seen something of this in the great C minor Étude, Op. 10, No. 12. And we have already noticed in the Mazurka, Op. 68, No.

[1] The vital *da capo* mark at the end of the second part of the C sharp minor Polonaise, Op. 26, No. 1, is missing in some editions. This has misled Leichtentritt (*Analyse der Chopin'sche Klavierwerke*) into condemnation of the 'not quite satisfactory binary form'. Leichtentritt's book, not to be confused with his earlier biographical *Friedrich Chopin*, contains far too many slips of this kind. Although the most detailed analysis of Chopin's works so far published, and a book which I myself have found very useful, its value is seriously diminished by innumerable errors in both text and music-type, I have found Leichtentritt's views on Chopin's harmony—views sometimes self-contradictory—of hardly any use at all.

3, and the posthumously published Valse in E minor the 'perspective foreshortening' of the return to the first section. This curtailing of the third section of the ABA form became the rule rather than the exception with Chopin, and it is interesting to note how the same proportion of third section is retained even in some of the most drastically curtailed instances; for instance, in the Prelude in B minor, Op. 28, No. 6, this third section consists of four out of the twenty-six bars of the whole piece, in the Mazurka in the same key, Op. 33, No. 4, of thirty-two bars out of two hundred and twenty-odd, in the C sharp minor Nocturne, Op. 27, No. 1, of eighteen bars out of just over a hundred: that is, in each case, of about a sixth of the whole piece. The same proportion is felt again in the C minor Polonaise, Op. 40, No. 2, which is in a sort of compound ternary form with the last two sections amputated: ABA CDC A.

Chopin often offers some sort of compensation for this curtailment; he does not trust solely to one's sense of perspective to restore the balance. In the Polonaise just mentioned he takes a little three-note motive, E flat, C, B flat (which may or may not have been derived from the principal theme of the piece, but which at any rate was utterly insignificant when it appeared before in a parallel passage), builds it up for four bars to a climactic point marked by the actual return of the first theme, and then accompanies that theme with a motive derived from the E flat-C-B flat idea. The object is not merely to conceal the join, which is done quite skilfully, but to give additional weight and breadth to an architectural section that has been robbed of a good deal of length.

In other cases, for instance the C sharp minor Nocturne and the D flat Prelude, the compensation takes the form of a new idea introduced in the coda: in the former, one of astonishing freshness and beauty. Chopin's codas are often memorable as perfect roundings-off of perfect poems, but, as here, they not infrequently have a structural function as well. (On the other hand, his introductions are seldom important structurally; they serve only to attract attention, or in his later works, more subtly, to place a harmonic curtain before the tonic key and so heighten the effect of its first appearance.)[1] In one piece, the B major Nocturne, Op. 32, No. 1, the extremely dramatic coda entirely takes the place of a repeat of the first section. At the end of the middle portion, which is actually related to the first by its stretto cadences, the music glides down to a dominant seventh which is succeeded, not by the expected tonic chord and opening theme, but by a foreign chord, drumlike throbs, and a tragic recitative: perhaps the most Wagnerian of all the numerous passages in which Chopin anticipates Wagner.

In the B flat Prelude, Op. 28, No. 21, the repeat

[1] See, for instance, the Mazurka, Op. 30, No. 4, the Scherzo in C sharp minor, and the Sonata in B flat minor, all written on the very eve of the third period. There were plenty of classical precedents for the use of this tonal curtain, but the Viennese masters rarely employed it with Chopin's subtlety.

is not drastically replaced by an entirely new element as in this nocturne; but only the chromatic accompaniment figure returns, now in its own independent right, rises to a climax (the peak of the whole prelude), dies down in a passage exactly parallel to one in the early part, but now entirely changed in significance, and spontaneously generates a new coda figure. The piece is one of the finest examples of Chopin's formal art on the tiny scale, for the rudimentary ternary form is completely transfigured.

Nor was his skill limited to this very miniature type of piece. Of all Chopin's smaller compositions the most sectionalized, next to the valses, are the mazurkas. Sharply contrasting sections are natural in dance-music and therefore perfectly justifiable in glorified dance-music such as these, and even when the composer relies on nothing more than the all-pervading mazurka characteristics to hold the piece together one is quite satisfied. But as early as Op. 24, No. 4, he begins to be no longer content with this. The piece is 'framed' by a twelve-bar introduction and thirty-two-bar coda; the main part of the mazurka grows out of the introduction so spontaneously that one hardly realizes that the opening, after the first four bars, is still only introduction, especially as the main theme of the mazurka is foreshadowed in bars 5 and 6; the return of the first section is contrived under the cloak of the capricious double-dotted figure which first appears four bars after the first double-bar; and the coda is based on the alto part of the introduction (bars 6-8). Again in Op. 30, No. 4, one of the first of those great mazurkas which are much less dances than tragic poems, we find a coda that not only looks back to the thematic introduction but combines

Ex. 19

the themes of both main and middle sections:

Ex. 20

With Op. 41, No. 1, an eve-of-third-period work, we already reach the almost symphonic style of the greatest mazurkas of the last phase. The return to the first section and the transition to the coda are both effected by quasi-development passages, so that the general outline appears as:

A *a*—16 bars.
b—16 bars.
B *c*—16 bars.
d—16 bars.
8 transition bars based on *a*.

A *a*—16 bars.
b—16 bars.
14 bars free continuation of *b*, building up to the—
Coda—8 bars of *a*.
13 bars based on the inversion of a motive common to *b* and *d*.

Even in more simply constructed mazurkas similar traits appear. Thus in Op. 41, No. 4, the greater part of the middle section consists of a miniature development of the opening theme.

Naturally, the more highly organized forms are in the minority, yet many of even the simplest of Chopin's maturer mazurkas and other short pieces are given some little touch that lightly disguises the naïveté of the underlying symmetry; the sections are repeated, as in Op. 24, No. 3, and the Nocturne Op. 27, No. 2, or the repeat takes a wrong turning and ends deliciously and unexpectedly, as in Op. 30, No. 1. The A flat Valse, Op. 34, No. 1, disguises its ternary form by a 32-bar 'refrain' common to each section—AX BX AX Coda—and the straggling shape of Op. 42, in the same key, is held together by the same simple device: AX BX CX DX AX CX.

But, except in the suite-like valses, Chopin's forms can seldom be accused of straggling. On the few occasions when he does break away from the fundamental ternary pattern it is nearly always into the nearly related ABACA form: the primitive rondo or scherzo-with-second-trio *à la Schumann* (cf. the Mazurkas, Op. 6, No. 1, Op. 7, Nos. 1 and 4; Op. 24, No. 2; the Prelude, Op. 28, No. 17, etc.). There are exceptions—the varied strophes of the E flat Nocturne, Op. 9, No. 2, the perfect 'arch' (ABCBA) of the Mazurka in E minor, Op. 41, No. 2, the strange, but emotionally inevitable, 'continuous' form of the G minor Nocturne, Op. 15, No. 3—but they are remarkably rare.

One other factor must be taken into account in considering even the smallest forms in Chopin: that of variable interpretation. A section may be repeated note for note as far as the printed music is concerned, yet, if Chopin's own practice counts for anything, the repetition was not necessarily intended to be exact in performance. With some composers—for instance, Beethoven—the printed notes are an attempt to represent music ideally conceived by the brain and inward ear; that ideal must have been practically immutable and it is the interpreter's duty to seek it through the printed notes. Chopin's contact with his sound-medium was more immediate; the printed forms of his works were often, as his own numerous variants and changes of mind show, the records of music that, however finely polished and worked out, was originally and essentially keyboard improvisation; the record remained but the improviser's own moods constantly changed. If Chopin did not go so far as Liszt, who actually improvised virtuosic trimmings to his interpretations, the spirit of improvisation was always present in his playing. A. J. Hipkins, the pioneer of Chopin-playing in

England, who heard the composer at Broadwood's in 1848, says: 'Chopin *never* played his own compositions twice alike, but varied each according to the mood of the moment, a mood that charmed by its very waywardness.'[1] We may safely assume that what was true of his interpretation of whole compositions was also true in a modified degree of his reading of repeated sections of a work. And this view is sometimes confirmed by the composer's markings. The first eight bars of the Étude, Op. 10, No. 9, for instance, have no fewer than five minor *crescendo* markings and the third bar is marked *con forza*; the note-for-note repeat which follows is marked at the beginning *sotto voce* and has only one *crescendo* marking. It is obvious that Chopin wished the second eight bars to sound like an echo of the first eight rather than like a mere repetition of them. And if this principle of variable interpretation plays a part in even the small forms, it must be far more important in some of the large-scale works, the scherzi and impromptus, which are after all essentially only small forms magnified and in which there are long stretches of note-for-note repetition.

Of these large-scale works two of the ballades, three of the scherzi and two of the impromptus (plus the Fantaisie-Impromptu) belong to Chopin's second period. The impromptus represent the lowest type of organism, the scherzi a slightly higher type, the ballades an altogether new and largely successful hybrid form, half lyrical and

[1] *How Chopin Played.* Dent. 1937.

akin to the short pieces, half epic and related to the principle of sonata form. And then of course there is also the B flat minor Sonata itself.

The Impromptu in A flat, the Fantaisie-Impromptu, and the B minor Scherzo are structurally of little interest. They are merely large-scale ternary structures with the joins in the carpentry ill-concealed. Even the effective return of the main section of the B minor Scherzo—the *fortissimo* opening chords breaking in among the dying echoes of the Polish Christmas song 'Lulajże Jezuniu' ('Sleep, baby Jesus'), which is the basis of the middle section—is very naïve by comparison with the best of Chopin's later transitions. The beautiful Impromptu in F sharp is also essentially in ternary form, though nine-tenths of the interest of the piece lies in the enormously extended and elaborated return of the first section: first in the 'wrong' key, a semitone too low, as in the finale of the E minor Concerto, then in the right one, but more and more completely veiled by variation.[1] In this piece, too, occurs a weak transition, perhaps the clumsiest in the whole of Chopin, despite the comparatively late date of the composition (1838)

[1] The beautiful opening melodic phrase, which Brahms must have unconsciously recollected a quarter of a century later when he set Ludwig Hölty's 'Die Mainacht', is itself a slight variant of a Ukrainian folk tune used by Rimsky-Korsakov in his opera *May Night*. In view of the close connexion between Polish and Ukrainian folk music, it seems probable that Chopin took his theme directly from a Polish melody. The point of similarity between Brahms's and Rimsky-Korsakov's 'May Nights' is a curious coincidence but nothing more.

and the attempt to make the link organic or at least thematic; the modulation from the D major 'middle' section (which is by no means at the structural centre of gravity) to the F major return of the first theme consists of two bars which, as Huneker says,[1] 'creak on their hinges'. They creak partly because Chopin, dropping into a harmonic habit which had by that time become almost second nature to him, reduced the progression to a mere skeleton, comprehensible enough to him because he mentally clothed it with flesh:

Ex. 21

but sounding very queer to ears that hear only the actual sounds; partly because even when thus flayed alive and unrecognizable, this particular chord-progression still does not fit the fragment of theme at all comfortably. The ingenious return to F sharp, twelve bars later, is as effective as this is the reverse.

Even the most enthusiastic admirers of Chopin would hardly claim that the F sharp Impromptu is a masterpiece of form. Or the G minor Ballade either. Yet the Ballade is recognizably a first and not unsuccessful experiment in a new and entirely

[1] *Chopin: The Man and his Music.* William Reeves. 1901.

individual form, a form that politely touches its hat to the superficial features of the classical 'first movement' but quietly ignores most of its underlying principles. The 'arch'-like outline of the G minor is simple enough:

Introduction: *Largo*.
First subject (*moderato*): G minor.
Second subject (*meno mosso*): E flat.
Development (*a tempo*): beginning in A minor and passing through various keys.
Second subject: E flat.
First subject (drastically shortened): G minor.
Coda (*presto con fuoco*): G minor.

The irregular recapitulation, with the second-subject themes returning before the first subject and still in the contrasting key, had a precedent in the first movement of another famous romantic work also composed in Paris and only five or six years earlier: Berlioz's *Symphonie fantastique*. Chopin notoriously disliked Berlioz's music, but it seems probable that he had no objection to taking a hint from it. His 'development section' in the G minor Ballade is even less Beethovenian than Berlioz's; it consists rather of variation and improvisation than of true development. That was inevitable with such lyrical themes. But it fulfils the same structural function, with its perambulation through strange keys—though even here Chopin blunders upon E flat long before he is ready to get back to the business of recapitulation.

And there are other signs that Chopin was beginning to understand something of the real essence of sonata-form. For instance, that recapitulation is not merely repetition, that if things have really happened in the development one cannot, without artistic falsehood, tamely return to the *status quo ante*. Consider the treatment of the *meno mosso* second subject here. The apparently insignificant florid link in the exposition (bar 8) has grown in the development and thrown its tendrils over four bars of the theme; and in the recapitulation it is the quintuplet-crotchet idea evolved from this which dominates the whole passage.

The Second Ballade, Op. 38, is a modification of the same form, and there is good reason to suppose that its original version may have been constructed on exactly the same lines. Schumann, to whom it is dedicated, wrote that 'its impassioned episodes seem to have been afterwards inserted. I recollect very well when Chopin played the Ballade here; it finished in F major, now it closes in A minor'. What Schumann meant by 'its impassioned episodes' is not altogether clear; perhaps the powerful sequential passage that crowns the *presto con fuoco* second subject, perhaps the more strenuous incidents in the working out of the first subject, which follows this, or perhaps the stormy coda. But the return of the first subject is foreshadowed in the bass octave passage just before the coda and does actually occur, with dramatic effect, as a fleeting wistful reminiscence when this coda has reached its climax. Imagine, then, that, instead of the coda, heralded and rounded off by fragments of the first subject, Chopin had here simply written out the first subject in full and in F major; we should have a form almost exactly like that of the G minor Ballade. And it seems to me highly probable that this or something like it *was* the original form of the piece that Schumann heard.

But if Chopin did originally end the piece in this way, he must have felt the artistic falsity of such a conclusion. (Apart altogether from the fact that the Ballade was admittedly 'inspired by some poems of Mickiewicz' which presumably demanded a tragic ending.[1]) Accordingly he altered it, even its key, and the result is effective and highly dramatic. But in contriving a dramatically satisfactory form he was unable to avoid battering the purely musical form rather badly and the marks of his tinkering remain visible.

From the structural point of view, the scherzi are much less interesting than the ballades, but

[1] As far as I can discover, Chopin himself never admitted anything more definite than this. But Huneker, in *Chopin: the Man and his Music*, first commends him as a 'wise artist' for giving no more definite clue, and then goes on, without quoting any authority, to speak of 'the G minor Ballade, after *Konrad Wallenrod*', to assert that the Second Ballade was 'written, Chopin admits, under the direct inspiration of Adam Mickiewicz's *Le Lac de Willis*', and to speak of the A flat Ballade as 'the *Undine* of Mickiewicz'. (On the F minor he says nothing, though there is a sort of legend connecting it with the poem *Trzech Budrysów*.) Now Huneker was cursed not only with an unduly florid prose style but with a tendency to intellectual pretentiousness that only too often arouses suspicion that he was something of a charlatan. If he had ever read *Konrad Wallenrod*, he would have known that

even in these the idea of sonata form is perceptible. Op. 31 has its two subjects in tonic minor and relative major, though this relationship is naturally unaltered in the repeat, and the trio proper is immediately followed by a quasi-development of it. And Op. 39 dispenses entirely with a trio, the middle section consisting first of passage-work, then of a variation of the chordal second subject with its delicious showers of *leggierissimo* quavers—in short, of 'development' as Chopin understood it.

Even in the first movement of the B flat minor Sonata the development is essentially an improvisation on the first subject, an agitated figure admirably adapted to this wild hunt through a thicket of keys. In other respects, the movement is concise and satisfying. The transition from the first subject to the strongly contrasted yet somehow related second is effected with a masterful laconicism that reminds one of Schubert's in the corresponding

there is about as much connexion between the G minor Ballade and Mickiewicz's grim narrative poem, with its exceedingly tough and unpleasant hero, as between *Macbeth* and Chopin's Barcarolle. As for *Le Lac de Willis* and *Undine*, I strongly suspect that they are one and the same poem: namely Mickiewicz's ballad, *Switezianka*, of which a translation is given, as *The Nixie*, in *Konrad Wallenrod, and other writings of Adam Mickiewicz*, translated by various authors and published in 1925 by the University of California Press. (On the other hand, by *Le Lac de Willis* Huneker may have meant *Switéz*, which is also a poem about a haunted lake.) *Switezianka*, which Rimsky-Korsakov long afterwards set as a cantata, might conceivably have inspired the Second Ballade, but it is absolutely preposterous to suggest that it has any connexion with the Third.

part of the Unfinished Symphony. And the curtailment of the reprise by the omission of the first subject was inevitable; to have brought it in again after worrying it to death in the pseudo-development would have been fatal.

Yet, fine as the movement is, it is something less than first-rate Chopin. For one thing, the whole piece consists of unmodified four- or eight-bar phrases (only the very last bar is thrown in as an extra) and such undisguised squareness of phrasing sustained for such a long period is unusual with Chopin. Symmetrical phrasing is the rule rather than the exception in the études, where it often usefully holds together what might easily have become an amorphous play of motives, and it inevitably predominates in the dance-forms and in the song-like nocturnes. But, especially in the latter, Chopin usually takes considerable pains to conceal the fact. Consider, for instance, how in the opening of the A flat Nocturne, Op. 32, No. 2, he evades the seemingly unavoidable squareness of a sustained B flat in the fourth bar of the main melody, extending it ornamentally instead and then gliding imperceptibly into the repeat. He adopts exactly the same procedure at the equivalent points in the D flat Prelude and the Nocturnes, Op. 27, No. 1, and Op. 37, No. 1, to name only three of the most obvious cases. And he has at his command a score of other devices for the same purpose. The *cantabile* melody of the Fantaisie-Impromptu, for instance, just contrives to avoid banality by the premature entry of the second four-bar clause, half a bar too soon—an example of what it is tempting to call '*rubato* of phrase conception'.

The same principle is often carried further, as in the middle section of the C sharp minor Polonaise, Op. 26, No. 1, where Chopin cuts across the 3/4 rhythm with a motive that suggests a contraction to 2/4, and then inserts a penultimate compensating bar:

Ex. 22

As Leichtentritt points out, the passage is a genius's version of what a mediocrity would have written as:

Ex. 23

Much of the capricious charm of Chopin's music is due to similar irregularities of phrasing or to the persistent use of unusual phrase-lengths, such as the six-bar units of the Mazurka, Op. 24, No. 3. He was a perfect master of the arts of phrase-overlapping, phrase-extension, and phrase-contraction. In at least one delightful instance, the end of the Mazurka in A flat, Op. 41, No. 4, a phrase is not merely contracted but deftly cut short. We hear only six of the anticipated eight bars and are neatly dropped into a silence as unexpected as that which terminates Act I of *Pelléas et Mélisande*. Chopin here does rhythmically what Debussy more than half a century later did harmonically. At the end of the E flat minor Étude, Op. 10, No. 6, on the other hand, he produces a charming effect with a thirteen-bar phrase produced by the insertion of five additional bars before the final chord.

Chopin's phrase-extensions are specially interesting, for their very nature is often peculiarly conditioned by the type of melody. Two examples must suffice: from the Mazurka in B minor, Op. 33, No. 4, and from the Nocturne in F, Op. 15, No. 1. The predominance of motive-generated melody in the mazurkas has already been mentioned; in this particular mazurka the first eight-bar phrase is extended to twelve bars by a process of generation from the melodic cell A-C sharp:

Ex. 24

A similar but less subtle play with a final motive is responsible for bars 21–2 of the same piece. In the Nocturne the melody is purely lyrical and the structural extension, the long yet uncompleted phrase that concludes the first part of the piece, arises from an essentially lyrical impulse to expansion. The effect is heightened by the irregular grouping of the clauses (5 + 3 bars) coming after a series of four-bar units.

Despite these numerous exceptions, of course, the four-bar or eight-bar phrase remains the foundation of Chopin's melody; the regularity is as essential to the charm of the irregularities as firm tempo is to the charm of tempo rubato. The metrical regularity is far less often broken than skilfully concealed. In this as in practically every other aspect of his melody, Chopin never completely shook off the two great primary influences that formed it: the contemporary Italian opera aria and the Polish folk-dance tune.

That Chopin's melody is heavily indebted to Bellini's is a commonplace of criticism. One has only to open at random the score of, say, *I Capuletti e i Montecchi* (produced in 1830), to be confronted with a melody that might easily belong to a Chopin nocturne:

Ex. 25

But the debt is not to Bellini only.[1] It is significant that as early as 1825 Chopin wrote 'a new polonaise on the *Barber*' (now lost) (letter of November, 1825), and the following year based the trio of a Polonaise in B flat minor on an air from *La gazza ladra*, that the Tarantelle of 1841 is obviously indebted to Rossini's, that Chopin could accompany from memory the once celebrated tenor, Franz Wild, in an air from Rossini's *Otello*. Nor did Chopin absorb all his Italianism directly. There can be little doubt that some of it was acquired at second-hand through Elsner and through his admired Spohr. Spohr's violin cantilena was admittedly based on Italian song and, as Dr. Maria Ottich points out,[2] 'does not consist merely of melody concealed by delicate tendrils of ornamentation but forms a self-contained line in which there is not a single merely decorative note'. Even the peculiar romantic *morbidezza* characteristic of Bellini and Chopin is common also to Spohr.

Still, it was the voice rather than the violin that Chopin admired, and it is even possible to detect in

Ex. 26

[1] Nor is the indebtedness all on one side. If Chopin owed much to his friend's melody, Bellini was not uninfluenced by the other's harmony. This is particularly noticeable in his *Puritani* (1835); see, for instance, Elvira's aria in Act II at the words 'Qui il giurava e poi crudele', and the remarkable chain of dominant sevenths in the duet in Act III.

[2] *Chopins Klavierornamentik*. Berlin. 1938.

his melodies purely vocal mannerisms, such as this effect of breath-taking in the E flat Nocturne (Ex. 26), which has dozens of parallels, many of them actually marked by brief rests (see Ex. 28*a*, end of bar 3).

All the same it would be a disastrous mistake to regard the melodies of the nocturnes and Chopin's nocturne-like melodies in general as mere echoes or imitations of Italian song-melody. As I have said elsewhere,[1] Chopin had 'an instinct amounting to genius for inventing melodies that would be actually ineffective if sung or played on an instrument capable of sustaining tone but which, picked out in percussive points of sound each beginning to die as soon as born, are enchanting and give an illusion of a singing that is often lovelier than singing itself'. In other words, Chopin's melody, in so far as it is Italian, is not an imitation but a stylisation of Italian *bel canto*.

To take only one example of his method: he produces marvellous pseudo-cantabile effects by repeating a note instead of sustaining it. (See the opening of the B flat minor Nocturne, Op. 9, No. 1, for one of hundreds of cases, bar 11 of the *sostenuto* section of the First Impromptu for a slightly more subtle example of melodic intensification by this means.) The device is in the direct line of descent from the *vibrato* and *Bebung* of the clavichordists; Mozart and Beethoven did not overlook it; but Chopin's peculiar application of it to this particular type of melody owes less to the practice of these great predecessors than to that of John Field (cf.

[1] *A Hundred Years of Music*. Duckworth. 1938.

the opening of the latter's Nocturne No. 2 in C minor). And we must not overlook the fact that the Italian opera-composers themselves employed both *parlando* effects—'speaking' on a single note—and coloratura note-repetition.

Not only do the melodies themselves remind one of Italian opera. Their treatment—the too seductive thirds and sixths of the D flat Nocturne, the Mazurka, Op. 17, No. 1, and a dozen other works—often seems to have been suggested by the same models. Chantavoine[1] even contends that Chopin based his famous *tempo rubato* on the practice of 'the tenors and sopranos of the Italian Opera who compelled the conductor to respect their "good notes"' and all the rest of the virtuoso singer's stock-in-trade. It is certainly true that the invention of *tempo rubato* has been attributed to an Italian singer-composer, F. A. M. Pistocchi (1659–1726),[2] but the practice was well known not only to the Italians but to Mozart and other South German composers of the eighteenth century, and (more to the point) was employed by Chopin's own master, Elsner, and other Polish composers.

Chantavoine is probably on safer ground when he suggests that Chopin's melodic ornamentation 'does not so much resemble the cadenzas of the classics and the fioriture of the German pianists—Moscheles, Hummel and the rest—as the vocal

[1] In the essay, 'L'Italienisme de Chopin', in *Musiciens et poètes* (Paris, 1912).

[2] See Lucjan Kamieński's essay, 'Zum tempo rubato', in the *Archiv fur Musikwissenschaft*, I, 1.

fireworks of the virtuosi of *bel canto*'. Hummel is not so easily dismissed as that. Musical genealogy is seldom even as simple as Solomon's little problem; any element in any composer may have as many as half a dozen parents, and Hummel's fioriture make him liable to one of the several affiliation orders that might be issued in respect of Chopin's. Nevertheless, the influence of vocal coloratura, perceptible in Field too, is unmistakable in Chopin from the *larghetto* of the F minor Concerto to the B major Nocturne, Op. 62, No. 1—though admittedly the patch of coloratura in the latter is exceptional in the later works.

But Chopin's is exquisitely stylized coloratura, not a mere imitation such as some of his predecessors had been content with—for, of course, he was not the first piano composer to take hints from the opera stage. 'Hitherto this species of adornment had only been modelled on the fioriture of the grand old school of Italian song', says Liszt.[1] 'Embellishments for the voice, although they had become stereotyped and had grown monotonous, had been servilely copied by the pianoforte; Chopin endowed them with a charm of novelty, surprise, and variety, quite unsuitable for the singer but in perfect keeping with the character of the instrument.'

Yet if Chopin stylized the methods of singers he studied them at first hand and, like C. P. E. Bach in his *Versuch über die wahre Art das Clavier zu spielen*, demanded that his pupils should do the

[1] *Life of Chopin*. (John Broadhouse's translation.) William Reeves, n.d.

same. And it is significant that he particularly admired fine coloratura. We find him praising Sabine Heinefetter for her 'coloratura like so many pearls', and 'enchanted' with Henriette Sonntag: 'Her *diminuendo* is the *non plus ultra* that can be heard; her *portamento* wonderfully fine; her chromatic scales especially toward the upper part of her voice unrivalled[1]. . . . She uses a few embroideries of a quite new type.' And he preferred Damoreau-Cinti to Malibran on the ground that 'the latter astonishes one, but Cinti charms. She sings the chromatic scales and coloratura almost more perfectly than the famous flute-player Tulou plays them'. The influence of pearly chromatic vocal coloratura is very obvious in many passages of Chopin (e.g. the Nocturne, Op. 9, No. 1, bar 3) and Dr. Ottich is surely right in regarding such effects as:

Ex. 27

(a) Nocturne, Op.9, No.3

[1] According to Niecks's translation. That by E. L. Voynich (*Chopin's Letters*. Desmond Harmsworth. 1932) is materially different: 'Especially her ascending chromatic scales are exquisite.'

as stylized *portamento*. Pianists would do well to bear that in mind.

On the other hand the quasi-recitative element in Chopin's melody—e.g. the left-hand octave passage in the C sharp minor Nocturne, or the end of the B major, Op. 32, No. 1—often seems to owe as little to the recitatives of contemporary opera as to the earlier instrumental recitatives of C. P. E. Bach and others. Recitative-like in intention and effect, these passages are essentially no different from a good deal of Chopin's purely ornamental coloratura; for they are as much cadenza as recitative and serve the same purpose of breaking the regular metrical pattern. One cannot imagine them with words fitted to them, as one can some of Liszt's instrumental recitatives.

Despite the enormous importance of chromaticism in Chopin's music, his melody is predominantly diatonic. In the vast majority of cases the chromatic elements in it are unessential and purely decorative: passing notes, parts of ornaments, and so on. (As we shall see presently, that is also true to a great extent of Chopin's harmony.) One is reminded of Abert's remark on 'the peculiar fusion of diatonicism and chromaticism in Mozart's piano style'.[1] 'Despite the intermingling of chromatic elements, the ground-structure of his melody is diatonic. . . . And the chromaticism serves far more often to smoothe off than to sharpen melodic contours.' That is almost equally true of Chopin, though there are a good many cases where he uses

[1] *W. A. Mozart*. Two volumes. Leipzig. 1924.

that typical nineteenth-century device of 'chromatic alteration', which has been aptly compared with the *musica ficta* of the Middle Ages, to intensify melodic as well as harmonic expression. The D natural in bar 5 of the C sharp minor Nocturne may be explained harmonically, but that at bar 13 is a different matter; the harmony calls for D sharp and the unexpected flattening is an intensification—which the composer has underlined with an accent mark. The melody strains down, as it were, toward the C sharp that follows; the leading-note principle here makes itself felt in the 'downward sharpening' (which we are paradoxically obliged to call flattening) of the second degree of the scale, just as centuries before it had made itself felt in the sharpening of the seventh degree of the modes. The same thing happens in bar 3 of the F sharp Nocturne, Op. 15, No. 2, which is a repetition doubly intensified: by the quintuplet replacing the group of four semiquavers in bar 1, and by the chromatic alteration of the fourth note of the quintuplet (in this case the sixth degree of the scale). On the other hand, the F double sharp in the melody of bar 5 is obviously conditioned by the harmony.

The sharpened fourths that occur so frequently in the mazurkas are another matter. As already explained, they are not 'alterations' at all but normal features of a scale commonly used in Polish folk music.

Polish elements also play a part in that rich ornamentation which is so peculiarly characteristic of Chopin's melodic style. The grace-notes and interpolated triplets characteristic of the folk mazurka, for instance, crop up continually not only in the mazurkas and polonaises, where one expects to find them, but in the valses (cf. Op. 69, No. 1, bars 1 and 3) and even in Chopin's most Italianate melodies (cf. the Nocturnes in D flat, Op. 27, No. 2, bars 11–13, 16, 21, etc., and A flat, Op. 32, No. 2, bar 9).

Arabesque melodic ornamentation in general is such an integral part of Chopin's music that one is tempted to propound the riddle, 'When is ornamentation not ornamentation?'—to which, of course, the answer is 'When it is Chopin's'. It may not be true of his early works, but his later compositions (the last most of all) confront us with the paradox of 'ornamentation' that constitutes the very substance of the thing ornamented. Even the straightforward scale and arpeggio passage-work of the classics is everywhere replaced by free arabesque lines.

Chopin inherited from his predecessors, from the harpsichord and clavichord composers and C. P. E. Bach to Mozart, Field, and Hummel, a rich collection of stereotyped ornaments—trills and chains of trills, *acciaccature*, turns, mordents, and the rest—all of which he drew on freely throughout his career.[1] And in common with his virtuoso

[1] In his mature works Chopin uses even these in an individual way. Note how he employs turns and inverted turns to emphasize repeated notes in the melodies of the D flat Prelude, the G minor Nocturne, Op. 37, No. 1, etc. There are classical precedents, yet

contemporaries—Thalberg and the rest—he enriched them by employing them, particularly the trill, in thirds and sixths. (In one third-period work, the in every way remarkable Polonaise-Fantaisie, he even wrote a three-part trill in thirds and sixths simultaneously.) These older forms of ornamentation are introduced most abundantly, perhaps superabundantly, in the earlier works: first, purely and simply as external decoration, then, even as early as the *Là ci darem* Variations and still more in the concertos, as expressions of creative exuberance.

With the concertos, however, at the same time Chopin begins to use ornamentation in the quite different, far more poetic spirit characteristic of his maturity. Like melodic chromaticism, one of its main functions is to veil and soften outlines that would otherwise be too clear and obvious—too classical. Coloratura decoration conceals the most commonplace harmonic cadences (Impromptu in A flat, bar 14 of the *sostenuto* middle section); less florid ornamentation serves to disguise note-repetition (Nocturne Op. 32, No. 2, bar 6). With Chopin ornamentation commonly takes the place of true variation, particularly in his less mature compositions. He is particularly fond of immediately repeating a simple *cantabile* phrase in a florid form:

in the classics such intensifying ornaments might be removed without very serious damage to the melody—they are, after all, only ornaments—whereas in Chopin they are indispensable parts of the melodic line.

Ex. 28

(a) Concerto, Op.21

(b) Nocturne, Op.55, No.1

Nothing could illustrate more strikingly the evolution of Chopin's use of ornament than this juxtaposition of the first solo theme of the F minor Concerto with a passage from the third-period Nocturne in the same key, written some fourteen years later. The ornamentation in Ex. 28*a* is pure coloratura, a typical example from the period when the exuberance of sheer virtuosity was being transmuted into wealth of poetic utterance. That in Ex. 28*b* is much more restrained, producing a variation not unworthy of Brahms himself. The distinction between inner substance and outward decoration is finally obliterated; both are fused into a new melodic line, a line which, like the instrumental lines in numerous suite and concerto movements of J. S. Bach and other composers of the baroque period, is as much significant 'passage' as melody in the everyday sense.

'Significant line', something between melody and passage-work, but originating in embroidery of a harmonic background, played an ever more and more important part in Chopin's music in what I have called his third period. But there are naturally a number of instances in late second-period works. The first section of the Impromptu in A flat, the finale of the B flat minor Sonata, the F minor Etude, Op. 25, No. 2, the Preludes in B major, E flat minor, and B flat minor—practically all eve-of-third-period works, dating from 1838–9—are symptomatic. From the familiar broken-chord technique of the Viennese classical period, Chopin here evolves infinitely more subtle effects through thinking in terms of more advanced, chromatically complicated harmony and by the free weaving in of passing notes, ornaments, and even ornaments-to-ornaments (e.g. *acciaccature* before notes that are themselves not true harmony notes). With Mozart and Hummel even the more complicated forms of broken chord, with chromatic apoggiature and other modifications, remain unmistakably harmonic and 'passage'-like. It remained for the mature Chopin to give this type of formation linear significance and thematic importance.

On the other hand, Chopin's trick of embedding a melody, as it were, in broken chords—producing what Hermann Abert has called 'singing figuration'—was mildly anticipated by Mozart (cf. the *più allegro* section of the C minor Fantasia, K. 475, and the first movement of the Concerto in the same key, K. 491). But he never used 'singing figuration' so poetically as Chopin does in the A flat Étude, Op. 25, No. 1, and the middle section of the B minor Scherzo. And the bringing out of a shadowy inner part in the upper notes of a widespread arpeggio accompaniment, as in the opening of the Étude in F minor, Op. 10, No. 9, is peculiarly Chopinesque.

It was in such details as this that Chopin improved on the superficially very similar accompaniment technique of Field. (Although Field himself was not altogether insensitive to the melodic implications of such accompaniment figures; at any rate he sometimes keeps one 'part' moving in thirds or sixths with the right-hand melody: cf. his Nocturnes No. 5 in B flat and No. 11 in E flat.) The Irish master used wide-spread left-hand arpeggios, usually in triplets, decorated them with chromatic passing notes and disposed them in more intricate shapes than the few simple formulae that had satisfied the classical writers—all of which strikingly enriched the sonority of the texture. In this respect, as in the poetic quality of his imagination, Field was emphatically superior to Hummel. But it never occurred to him to give this arpeggio accompaniment anything *more* than harmonic significance as Chopin did.

Nevertheless Chopin's debt to Field, not only in this aspect of his keyboard writing but in a score of others, was enormous.[1] Even in such a mature work as the Nocturne, Op. 32, No. 2, dating from

[1] 'Liszt's statement that the Nocturnes of Field were regarded by Chopin as "insuffisants" seems to me disproved by unexceptionable evidence', says Niecks. The weight of evidence is, indeed, all on the other side: that Chopin frankly admitted his admiration for Field's nocturnes and concertos, and constantly gave them to his pupils. But even if he had never expressed his opinion in words, it is expressed sufficiently in his own nocturnes.

1837, he was still capable of echoing the older man quite closely:

Ex. 29

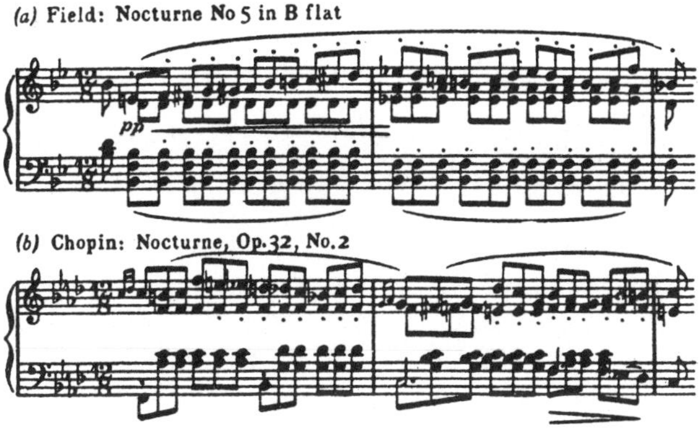

The similarity lies not in the notes—the sort of superficial resemblance that Sir Donald Tovey has ridiculed as 'top-knottery'[1]—but in the general idea, used in each case as a contrasting middle section in a purely cantabile piece.

The author of the article 'Pianoforte-Playing' in *Grove*, however, draws attention to an important difference between Field's piano-writing and Chopin's: 'the passages of embroidery in which both delight require, in Field's case, that perfect equality of finger at which the earlier writers aimed; in Chopin the essential weakness of the human hand is turned to beautiful account, for his passages are

[1] See *Essays in Musical Analysis*, Vol. i. Essay on Haydn's Symphony No. 88, in G.

often devised in such a manner that the weak finger of the hand has to play the note which is to be comparatively unimportant. In the concertos and studies the natural conformation of the hand is kept before the composer's eye, and, as a consequence, his difficulties are always of a kind that is grateful to the player, however intricate they may sound.' This principle of turning to account the inherent limitations of the human hand was carried still farther by Liszt.

And if Chopin learned from Field everything that the latter could teach him, he invented dozens of keyboard effects the older composer had never thought of: hazy, almost impressionistic pedal effects, contrasts of scoring both subtle (Étude in E, Op. 10, No. 3, bars 15–18) and striking (the D flat subject of the Third Scherzo, with its delicious *leggierissimo* showers of quavers interspersed among the *sostenuto* chord passages), left-hand bravura (Étude in C sharp minor, Op. 25, No. 7), persistent cross-rhythms (two of the Trois Nouvelles Études). Indeed he possessed a fertility in the invention of pianistic devices so inexhaustible that, a few favourite formulae apart, hardly any two pieces seem to be laid out in the same way.

In all this, of course, one sees the exuberance of the virtuoso working with absolute ease in his own peculiar medium. If the inspiration did not actually come first from the fingers, at least it must have come from fingers and brain simultaneously working in closest collaboration. It is impossible to hear such pieces as the Preludes in F sharp minor and F major, and not to realize the creative impulse that Chopin drew from the very feeling of his hands on the keyboard—the keyboard, be it remembered, not of our modern concert grand with its heavy touch and iron resonance, but of the old-fashioned 'square' piano with its simpler mechanism, lighter touch, and thinner tone.

There can be little doubt, I think, that Chopin's harmony—the most important, most individual, and most fascinating of all aspects of his music—was also largely inspired, or at any rate discovered, in the same way: by improvisation at the keyboard. There may have been precedents for some of his harmonic exploits, notably in Spohr, but it is obvious that many of them were directly inspired by the timbre of the instrument or brought to light by the improviser's delicate fingers. And this is all the more important since even Chopin's basic ideas are frequently harmonic rather than melodic. The already mentioned case of the second section of the posthumously published Valse in E minor[1] is but one of many.

Now although the charm of Chopin's harmony lies almost entirely in its iridescent chromaticism, it is, like his melody, fundamentally diatonic. The diatonic basis is usually chromatically embroidered, frequently even temporarily broken up altogether. But its presence is always felt. (And, of course, there are a good many pages of Chopin as uncoloured by chromatic harmony as a page of typical Haydn.)

The chords that form the basis of Chopin's

[1] See p. 28.

exceptionally rich harmonic vocabulary are simple and comparatively few in number: the ordinary diatonic triads and inversions, the major triads on the flattened supertonic (=Neapolitan sixth)[1] and flattened submediant, the minor form of the subdominant triad in the major mode, a favourite chord with all the composers of the romantic period (e.g. in the Mazurkas, Op. 7, No. 4 and Op. 33, Nos. 2 and 4), the dominant seventh (even the dominant ninth is comparatively uncommon), the diminished seventh, the added sixth (including its minor form in the major mode), and the so-called German sixth on both flattened sixth and flattened second degrees of the scale. The other—the 'French'—form of augmented sixth is rather rare in Chopin: he uses it for the peak chord at the end of the F major Ballade, as if feeling it to be peculiarly intense. Secondary sevenths and ninths seldom or never occur, unless we consider the added sixth an inversion of a secondary seventh. The secondary ninth twelve bars before the end of the A flat Ballade, a third-period work, is quite exceptional.[2]

[1] I must remind the reader that in speaking of chords I have in most cases not differentiated between inversions and root-positions, as the tonal functions are usually the same whatever the inversion.

[2] The numerous chords foreign to the key with which Chopin's music is liberally sown may be understood in the light of the German theory of interdominants (*Zwischendominanten*), which will be explained later, better than in the twilight of the lamentably misnamed Dr. Day's theory of supertonic and tonic 'fundamental discords' which has cursed English harmony-teaching since the middle of the last century.

But Chopin often overlays these simple basic chords with appoggiaturas, suspensions, anticipations, and passing notes so richly that, especially in his later works, they are almost unrecognizable. And analysis of his harmony is by no means simplified by his habits of presenting progressions in the sketchiest of skeleton forms, as in the already mentioned modulation in the F sharp Impromptu,[1] the opening of the C sharp minor Scherzo, and the finale of the B flat minor Sonata, and of writing in convenient rather than grammatically correct notation.

Even in such comparatively early works as the E flat Nocturne and the 'Black Key' Étude we find Chopin using suspensions and passing notes with characteristic freedom. In the second half of the very first bar of the Nocturne a suspension (F in the right hand) and a passing note (D in the bass) combine to prevent the E flat chord from being heard at all in its pure form. And the Étude contains striking examples of the effects a genius can wring from a single chord sustained for four bars (25–8) with diatonic passing notes in the treble and chromatic passing notes in the bass, and, five bars later, from a suspension (the D flat) held for four bars—or rather scattered over three octaves for four bars—implied in the fifth and resolved only in the sixth. The distribution of the suspended note by figuration is, of course, thoroughly characteristic of Chopin. Writing for the keyboard with its partial obliteration of true part-writing, he

[1] See p. 54.

felt perfectly free to resolve dissonant notes in 'wrong' parts or to introduce passing notes in parts other than those to which they rightly belong.

Of the innumerable beautiful or powerful transition chords (or pseudo-chords) produced by suspensions or passing notes, two examples must suffice:

Ex. 30

(a) Nocturne, Op.15, No.1

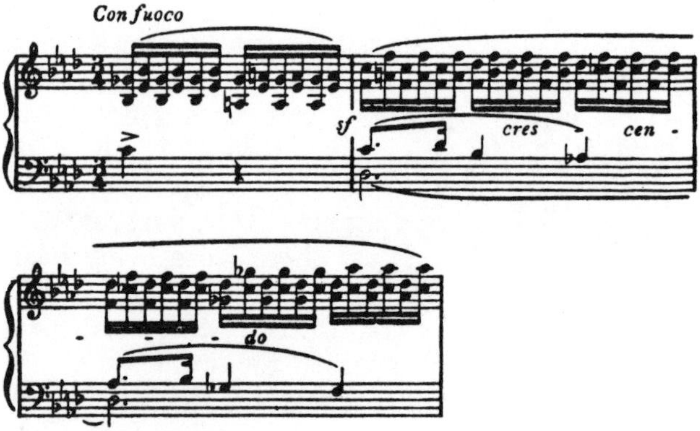

(b) Nocturne, Op.27, No.1

The faint clash of the suspended C and A natural in the second bar of Ex. 30*a* is as delicious as the G flat passing notes in the third. (It will be noticed that the true harmony does not change at all in bar 3.) In Ex. 30*b* the chromatic passing note produces a pseudo-chord of great intensity.

From this it was but a step to the use of such formations as chords in their own right, omitting the 'true' note altogether: the introduction in harmony of the same principle of chromatic alteration which we have already observed in Chopin's melody, and which was ultimately to lead nineteenth-century harmony through *Tristan* to the Schönberg of the *Klavierstücke*, Op. 11. Indeed the so-called 'French sixth' on the flattened supertonic, which Chopin himself used from time to time (e.g. Prelude in C minor, bar 6, and Polonaise in A, eighteen bars before the end of the middle section), is really the dominant seventh (second inversion) with the fifth chromatically altered.

This multiplication of appoggiaturas, passing notes, suspensions, and anticipations, usually in chromatic forms, produces an extraordinary plasticity, even fluidity, in the harmonic substance, so that at times *all* the parts are found to be moving by semitones as in *Tristan*. Transitional passages produced simply by semitonal movement of parts are very common in Chopin. Sometimes the movement is quite obvious, as in the E minor Prelude, the opening bars of the B flat minor Sonata and the Mazurka in the same key, Op. 24, No. 4, the Mazurka, Op. 17, No. 2 (end of middle section);

sometimes exquisitely concealed by the keyboard lay-out, as in the D flat Nocturne, Op. 27, No. 2 (bars 42–5). And Chopin uses the same technical device for very different aesthetic purposes: in the Sonata and the B flat minor Mazurka to conceal the tonic key and enhance the effect of its appearance when it does come;[1] in the E minor Mazurka to modulate, in the Prelude to suggest profound and deepening gloom, in the Nocturne to add a new range of shimmering harmonic colours to the already rich key of D flat.

The last example is perhaps the most characteristic and the most important. Any bungler could have seen what possibilities the device offers for easy modulation. Hundreds of extemporizing organists have done so, if with something less than Chopin's skill. Spohr had discovered its uses as a means of extending a key or even temporarily obliterating tonality. But it remained for Chopin to use it in this sense with consummate artistry, subtly disguising it, avoiding the dangers of its merely meandering tendency, giving it purpose and direction, and drawing from it its maximum possibilities of colour.

Perhaps another and more virile example, taken from the Étude, Op. 10, No. 8, will make this clearer:

[1] There is a Mozartian precedent just before the reprise in the first movement of the E flat Piano Concerto, K. 449.

Ex. 31

It will be seen that this apparently fairly complicated progression really consists of this very simple one:

Ex. 32

—that in fact there is no modulation at all. The music proceeds from one inversion of the dominant seventh to another simply by semitonal movement of parts, the intervening chords being purely transitional and having no tonal function. The chord at bar 4 is neither the dominant seventh of E flat it appears to be nor even a German sixth in D minor written enharmonically, but simply an accidental transition form. (There are dozens of similar cases in Chopin; for instance, the chord in the second bar of Ex. 21.) The whole of Ex. 31, then, is not a modulation but an extension of D minor, 'an iridescent play of colour, an effect of superficies, not an effect of substance', as Hadow said of the similar but much briefer and more delicate passage in the E flat Nocturne.[1] At the same time it is obvious that its D minor-ness is determined only by its first and last chords; removed from its context, the passage would be completely atonal.

It will be noticed that no fewer than three of the transition chords in Ex. 31 take the form of diminished sevenths (as do a great many of Spohr's transition chords). Indeed the diminished seventh plays an extraordinarily important part in the undermining of tonality not only in Spohr and Chopin but in their immediate successors, Liszt and Wagner.

We have already noticed the semitonal side-slip of diminished sevenths in Chopin's early works (see Ex. 9), ornamental passages temporarily breaking up the tonal surface. Naturally this idea is elaborated in more mature compositions. Sometimes it remains decorative, if more skilfully laid out, as in the middle section of the E major Étude, Op. 10, No. 3. But already in another of the Op.

[1] See p. vii.

10 Etudes (No. 1 in C) this progression ceases to be mere glittering decoration and becomes colourful, non-tonal substance (bars 8–6 before the end). As such, it lends itself equally to exquisite melting effects, as in the Nocturne in D flat, Op. 27, No. 2, bars 62–4 and 66–8, where the tonic pedal prevents its losing its tonal bearings, and to the expression of intense excitement, as in the peroration of the F major Ballade. In the B flat minor Scherzo (bars 17–22 of the final *Più mosso*) it is used in exactly the same sense as in the Ballade, but intensified by appoggiaturas. In the Mazurka, Op. 7, No. 2, the succession of sevenths is further disguised by suspensions producing intermediate transition chords:

Ex. 33

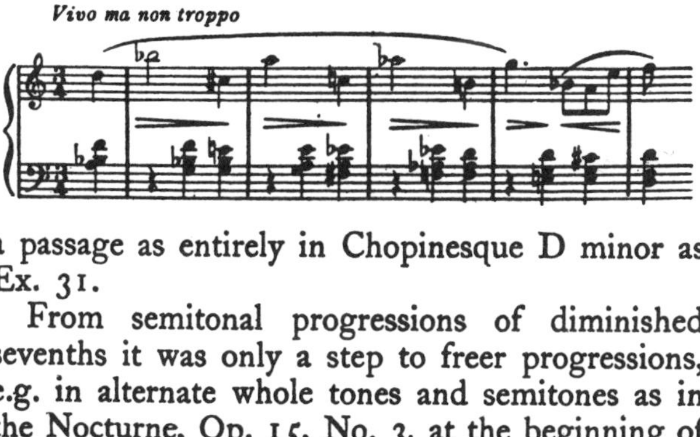

a passage as entirely in Chopinesque D minor as Ex. 31.

From semitonal progressions of diminished sevenths it was only a step to freer progressions, e.g. in alternate whole tones and semitones as in the Nocturne, Op. 15, No. 3, at the beginning of the transition to the second section. And from the chromatic side-slipping of diminished sevenths only another step to the chromatic side-slipping of dominant sevenths; Chopin takes it in the C sharp minor Scherzo, the Mazurka in the same key, Op. 30, No. 4 (a curiously Debussyesque passage), the A flat Nocturne, Op. 32, No. 2, and other works. The two following examples are specially characteristic of Chopin's appoggiatura and figuration technique, and of his love of piquant sound-clashes:

Ex. 34

(*a*) **Prelude, Op. 28, No. 8**

(*b*) **Etude in D flat (from 'Trois Nouvelles Etudes')**

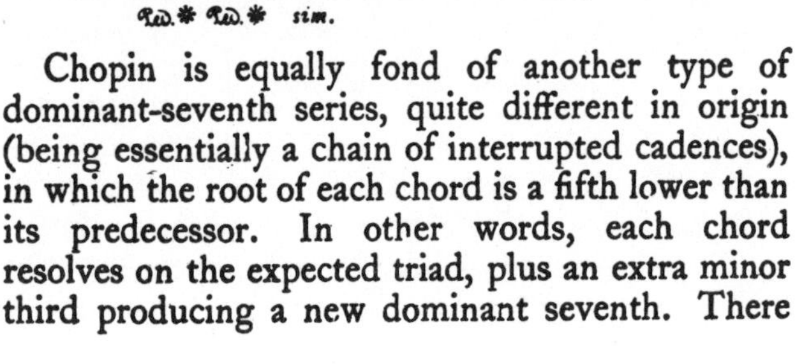

Chopin is equally fond of another type of dominant-seventh series, quite different in origin (being essentially a chain of interrupted cadences), in which the root of each chord is a fifth lower than its predecessor. In other words, each chord resolves on the expected triad, plus an extra minor third producing a new dominant seventh. There are numerous examples in Chopin's work. Thus in the Étude, Op. 10, No. 1 (bars 24–35), he leaps from a dominant chord in A through a series of dominant sevenths (disguised by suspensions and appoggiaturas) in D, G, C, F, and B flat to E flat. The F in the last chord is then flattened, producing enharmonically a French sixth (altered form of dominant seventh) in A. The A major triad appears in the next bar, after a ten-bar parenthesis. At the end of the C sharp minor Étude, Op. 25, No. 7, a much 'tighter' chain of sevenths is given a characteristic twist at the end, producing a brief parenthetic emphasis on the key of the dominant before dropping back into the tonic:

Ex. 35

The first chord of the second bar, instead of the anticipated dominant seventh in A, is merely a transition form (the D sharp being suspended instead of resolving on to D natural) leading to the minor form of the added sixth in G sharp. But the next chord, the dominant seventh in G sharp, resolves on to an added sixth in C sharp minor—and we are home again. We thus get a sort of full close raised to the second power.

Other interesting examples of such chains of sevenths will be found in the Polonaise, Op. 26, No. 1 (bars 13–14 of the *meno mosso* section), and in the Mazurkas, Op. 6, Nos. 1 (bars 5–9) and 3 (bars 27–8 of the middle section). In the two latter cases the chain is characteristically disguised by passing notes producing transition chords. In the Mazurka, Op. 17, No. 1, the chain is dislocated in the middle by a foreign chord, D–F–C–A flat, taken as a chromatically altered form of the chord on D flat and left as an altered form of the chord on D natural. But the dislocation is characteristic of Chopin. If he has mannerisms, he never actually repeats himself; some new disguise, some unexpected twist always gives freshness and vitality to what might otherwise have degenerated into mere formula.

These chains of dominant sevenths are, as I have said, elaborations of the interrupted cadence. Extremely lavish employment of interrupted cadences is one of the most important symptoms of romantic harmony in general and Wagner's in particular, and Chopin also anticipates Wagner in other favourite forms of it, such as resolution on a chord with root a third lower than the expected one (e.g. Scherzo in B flat minor: bar 49 of the working-out section). Again, the dramatic interrupted cadence near the end of the Nocturne, Op. 32, No. 1, is extraordinarily Wagnerian.

These series of diminished and dominant sevenths show how much importance Chopin attached to harmonic progressions for their own sake, to what we may call the *absolute* value of such progressions. It was only to be expected therefore that he should have made very free use of series of absolute progressions: in other words, of harmonic sequences. Here again, quite as much as in his interrupted cadences and his free use of chromatic appoggiaturas and passing notes, Chopin was one of the direct precursors of Wagner. (He in turn, of course, must have been familiar with numerous precedents in earlier masters, above all in Beethoven.) And here, too, although he works the basic idea almost to death, his disguises and modifications of it are infinitely varied.

The harmonic sequence by whole tones, whether rising or falling, is the general rule with Chopin, that by semitones in the A flat Prelude (bars 51–3) being quite exceptional. A very simple, yet typical instance of harmonic sequence by whole tones occurs in the D major middle section of the Polonaise in A, Op. 40, No. 1. A cadence in the dominant is followed by a German sixth also in A major; this is taken in the sense of a dominant seventh in B flat and initiates a sequence of full closes in B flat, C, and D. The whole passage is yet another example of key-extension; there has been no real modulation at all. One suspects it was such passages as this which provoked the remark of Moscheles about 'the harsh modulations which strike me disagreeably when I am playing Chopin's compositions' and which annoyed more conservative contemporaries still more deeply.

Naturally Chopin's harmonic sequences are not

always simply extensions of a key. They are also a very common means of modulation, though seldom employed crudely without some touch of ingenuity at the crucial point of arrival in thc new key: a dominant seventh-German sixth or other enharmonic pun, a chromatic alteration turning a dominant into a diminished seventh or vice versa, in one case (the Étude, Op. 10, No. 4, bars 41–4) a dissolution into a shower of diminished sevenths before crystallization into the new key. Once again it cannot be too strongly emphasized that if Chopin had favourite devices he never, or very seldom, lapsed into routine. The more clearly one recognizes his basic formulae, the greater is one's delight in his variations of them. He sometimes repeats one or more members of a sequence (e.g. the G minor Ballade, bars 16–13 before the reprise); in the Mazurka, Op. 24, No. 3, he not only repeats one member but carries the harmonic sequence across the beat, just as he carries the melody across the beat in the Valse, Op. 42, and the E major Scherzo.

Like the side-slipped-diminished-seventh device, the harmonic sequence is used in Chopin's hands for very different aesthetic purposes: for effects of the most delicate poetry (Nocturne, Op. 37, No. 2, bars 25–7) and for powerful effects of glowing climax, as in the F major Ballade (bars 17–22 of the first *Presto*: a passage standing midway between the development section in the Adagio of Beethoven's *Hammerklavier* Sonata and the sequence technique of Liszt and Wagner).

Somewhat akin in feeling to the sequence are also those abrupt transitions to keys a tone, a semitone, or a third away which one finds so frequently in Chopin and other romantic composers. It is true there were precedents in the classics for these juxtapositions of distant keys. 'Mediant key-relationship' can be found not only in late Haydn and in the *Emperor* Concerto but in compositions of the Mannheim school. Haydn wrote a piano sonata in E flat with slow movement in E major; indeed semitonal relationship is recognized in the admission of the Neapolitan sixth and made easier by the dominant seventh-German sixth pun. But there is a vast difference between the occasional abnormal introduction of such effects and their employment on almost every page as normal parts of a composer's vocabulary. And an equally vast difference between the prosaic and the poetic.

Spohr knew all about 'tonal parenthesis'. There is a good example at the end of the Andante (a set of variations on Handel's 'Harmonious Blacksmith' theme) of his Octet, Op. 32, which Chopin heard in 1829 and thought 'lovely, exquisite':

Ex. 36

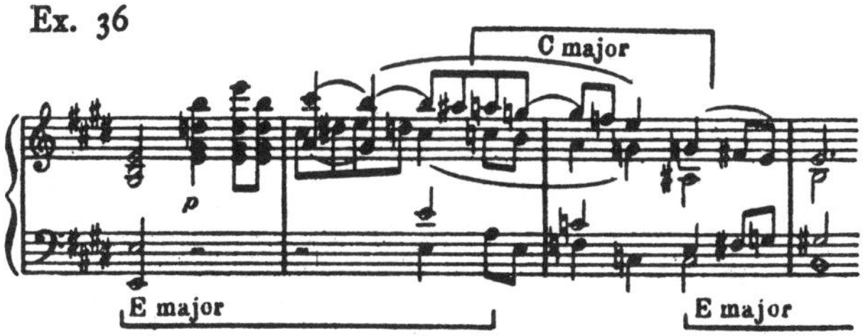

But how dry and unimaginative that sounds by comparison with these examples from two Chopin polonaises:

Ex. 37

(a) Op. 26, No. 1

(b) Op. 40, No. 2

Here in each case, as in the Spohr, the parenthesis arises from a simple enharmonic pun. But Spohr's is prose, Chopin's poetry.

We have still left untouched, however, one of the most fruitful of all sources of key-parenthesis or key-extension, one of the foundation-stones of chromatic harmony in general: the interdominant chords (*Zwischendominanten*, as German theorists call them) freely employed, not only by the Viennese classic masters, but by J. S. Bach and even earlier composers. Observing that all these masters introduced chords foreign to a key, the Father of English Harmony-Teaching, Alfred Day, tried to reduce their practice to theory by inventing a hypothetic system of tonic and supertonic discords parallel with the dominant series in each key. In so far as he recognized that chords foreign to a key do not necessarily produce even momentary modulation, Day was quite right; but in attempting to bring them into the fold by his hypothesis of supertonic and tonic 'fundamental discords'[1] he was wildly and hopelessly wrong. Even by the time his treatise was published (in 1845), it was insufficient to explain the practice of contemporary composers, notwithstanding which it provided the basis of English teaching of harmony for very many years.

For the complicated and unreal deductions of Day, it is necessary only to substitute one simple observation: that any chord in the key may be preceded by a foreign chord standing in dominant (or even subdominant) relation to it, whether dominant seventh or ninth, diminished seventh (perhaps the commonest form of interdominant), or simple triad. Thus in the B flat minor Nocturne, Op. 9, No. 1, the chords in the first halves of bars 13 and 14 are interdominants resolving on the subdominant triad and dominant seventh respectively,

[1] See p. 78.

and the second chord of bar 35, the dominant seventh on the submediant (which Day's theory does not account for at all), is an equally simple interdominant to the supertonic chord in the next bar.

The possibilities of the interdominant for key-extension or key-parenthesis—to say nothing of unexpected modulation—are almost unlimited. Obviously, each interdominant chord is itself a parenthesis within the key, or an extension of the key on a small scale, and Chopin's music is full of examples of interdominants introducing chains of dominant sevenths (e.g. the Polonaise, Op. 26, No. 1, bar 13 of the *meno mosso* section) or other tonal digressions. The frequency of interdominants in itself prepares the ear to accept without question the appearance of foreign chords in a key. Sometimes, instead of flying off at a tangent, Chopin omits the expected resolution of the interdominant altogether; the famous penultimate bar of the F major Prelude contains an interdominant to the subdominant triad—which is then omitted.

As an example of colourful key-extension the following series of interdominant chords in quick tempo, some of them resolving on chords which are themselves chromatic extensions of the key (triad on the flattened sixth and major triad on the supertonic), would be difficult to beat; it is taken from the Étude, Op. 10, No. 7:

Ex. 38

Like the opening bars of the same Étude it is completely in C major, but in a Chopinesque C major that differs very remarkably from the C major of the Viennese classics.

Such passages in Chopin are seldom modulatory. What appear to be rapid series of modulations in his music are seldom true modulations—the first page of the G major Nocturne, Op. 37, No. 2, and the C sharp minor Prelude of 1841 are among the rare exceptions—but rather extensions of key, with consequent weakening and undermining of tonality in the classical sense, in the direction followed by Wagner, Liszt, Franck, and Schönberg. Even the Prelude to *Tristan* hardly avoids the tonic more persistently than Chopin's Prelude in the same key, Op. 28, No. 2. And both establish a subconsciousness of the tonic key by the same means: a sort of polarization in dominant and sub-dominant regions. Similarly, the opening page of the Op. 45 Prelude, after the first five bars, is surely in E major—suggested by a balance of B major against A major even more than by the later appearance of E major itself.

III

THE LAST PHASE

(1841–1849)

THE compositions of Chopin's last and greatest period are:

Allegro de Concert, Op. 46 (1841)
Ballades in A flat, Op. 47 (1841), and F minor, Op. 52 (1842)
Barcarolle, Op. 60 (1846)
Berceuse, Op. 57 (1843)
Fantaisie in F minor, Op. 49 (1841)
Impromptu in G flat, Op. 51 (1842)
Mazurkas, Op. 50 (1841), Op. 56 (1843), Op. 59 (1845), Op. 67, No. 4 (1846), Op. 63 (1846), Op. 67, No. 2 and Op. 68, No. 4 (1849)
Nocturnes, Op. 48 (1841), Op. 55 (1843), and Op. 62 (1846)
Polonaises, Op. 44 (1841) and Op. 53 (1842)
Polonaise-Fantaisie, Op. 61 (1846)
Prelude in C sharp minor, Op. 45 (1841)
Scherzo in E, Op. 54 (1842)
Sonata in B minor, Op. 58 (1844)
Tarantelle, Op. 43 (1841)
Valses, Op. 70, No. 2 (1842), and Op. 64 (1847)

Practically everything that has been said of the second-period works applies equally to these, too.

To analyse them in the same way and at the same length would only be to cover much of the same ground. All that remains now is to show in what way even the masterpieces of the second period are surpassed by a number of those of the third.

Anyone who knows his Chopin must be struck on reading this list of last-period compositions by the number of works to which one inevitably applies the adjective 'great': the Fourth Ballade, the F minor Fantaisie, the Mazurkas Op. 50, No. 3, and Op. 56, No. 3, both the Polonaises, the Polonaise-Fantaisie, the Nocturne, Op. 48, No. 1, and the Berceuse and Barcarolle, which are surely the two greatest of the nocturnes. It needs no analysis of style to discover that in these works Chopin entered a new, wider, and loftier sphere, acquired an almost epic breadth of conception. But analysis will show us how, while the elements of the third-period style are practically all to be found in the compositions of the second period, Chopin now not only subtilized and refined them but added a fresh element by his newly acquired grasp of counterpoint.

Harmonically the compositions of the third period are distinguished above all by their subtlety and by the accumulation and elaboration of devices already familiar. The part-writing is freer, so that we find, for instance, anticipations of notes implied but not actually sounded in the true chord that follows (e.g. the D sharp in bar 22 of the E major Scherzo). Having accustomed himself to successions of dominant sevenths a semitone apart and a

Ex. 39

(a) Impromptu, Op.51

(b) Polonaise Fantaisie

(c) Ballade in A flat

(*d*) **Mazurka, Op.59, No.3**

fifth apart, Chopin now writes them in whole-tone successions, beautifully disguised by appoggiaturas, suspensions, and so on (e.g. the Mazurka, Op. 59, No. 3, bars 65–7, and the F minor Ballade, bars 72–5—the latter containing characteristic examples of transition chords incidentally forming diminished sevenths). Even the simplest harmonic bases are completely overlaid with wealth of detail. Consider these examples of elaboration of (*a*) a single dominant seventh, (*b*) a single diminished seventh (itself only an interdominant) over a dominant pedal, and (*c* and *d*) side-slipped diminished sevenths. (See Ex. 39.) Comparison of Ex. 39*c* with the harmonically similar passage at the final climax of the F major Ballade will show how Chopin had progressed in the plastic handling of his material. Ex. 39*d* is a striking anticipation of Wagner.[1]

[1] The direct influence of Chopin is apparent in a good deal of later music, outside the main stream of Liszt and Wagner. Drawing attention to the opening of the Mazurka, Op. 63, No. 3, as the probable origin of melodic motives in the prologue to *Snegurochka* and in Act III, Scene 2, of *Mlada*, Rimsky-Korsakov remarked to Yastrebtsev that 'in Chopin you will find many of the real roots of contemporary music'. (V. V. Yastrebtsev: *My Reminiscences of Rimsky-Korsakov*, entry of August 30th, 1894.)

The Mazurka, Op. 56, No. 3, contains a still more remarkable premonition of a typical Wagnerian procedure: quasi-contrapuntal treatment of a theme, condensed into a given harmonic framework—in this case again a chromatic succession of diminished sevenths:

Ex. 40

Though very different from true polyphony, this passage is thoroughly typical of that breaking up of vertical harmony into horizontal thematic lines characteristic of Wagner's mature style. Just as in the seventeenth century the old fluid contrapuntal technique already began to solidify into the chordal homophony that was the rule in the last half of the eighteenth century and the first half (if not more) of the nineteenth, so in Wagner this temporarily all-predominant homophony began again gradually to dissolve into linear elements—a tendency that led eventually, through Strauss and early Schönberg (*Ein Heldenleben* and the *Gurrelieder*) to the 'linear counterpoint' of Hindemith, the polytonality of Milhaud, the canonic complications of *Pierrot Lunaire*, and the twelve-tone system. And Ex. 40 is one of the first symptoms of this tendency.

Chopin now treats non-tonal chords with even

greater freedom than before, as chords in their own right, as it were (cf. the chord at bar 5 of the Polonaise, Op. 53). Interdominants are thickly strewn in the quickest passage-work (bars 15–14 before the end of the Polonaise, Op. 44) or lead to the boldest key-extensions (e.g. in the Nocturne, Op. 55, No. 2, bars 16–13 before the end, where the interdominant principle is carried to a higher power). Tonal parenthesis becomes extremely common and even occurs in quick passage-work (e.g. the staccato quaver chords just before the coda of the F minor Ballade).

On the other hand, sequences are now frequently used on a larger scale. Instead of being treated simply as extensions of the main key, they are now used more effectively as substitutes for development, very much as Liszt used them later, and less effectively, in his symphonic poems. The development section of the A flat Ballade, for instance, ends with a modified sequence: 8 bars in B, 8 bars in C, 4 bars in D, 4 bars in E flat. The compression and consequent shortening of the later members of the sequence, and their organic overlapping, again anticipate Wagner. The development section of the F minor Ballade, though not built up of actual sequences, is constructed largely on the same principle.

Chopin's new harmonic language, carried as far as he ventured to take it, may be studied in the A flat Mazurka, Op. 59, No. 2 (bars 81–8), in the F minor Mazurka, Op. 68, No. 4, said to be his last composition, and the Polonaise-Fantaisie, bars

231–54, the most remarkable passage in a work filled from beginning to end with harmonic intimations of Wagner. The pathetically sketchy F minor Mazurka, a sort of harmonic skeleton, is a useful reminder that advanced style is not necessarily synonymous with increased aesthetic value.

The keyboard layout of the last compositions shows an advance similar to that in the harmony. On the one hand we find a refinement of the already refined; the Berceuse, for instance, quite uninteresting harmonically, is one of the supreme examples of Chopin's art of keyboard *facture*—a perfect embodiment of all the outstanding features of the new style of piano-writing: 'percussive-singing' melody, use of the pedal, chromatic filigree-work over a diatonic foundation, contrast of registers, and so on. On the other hand, one notices a new strength and additional interest supplied by the fresh contrapuntal element, presumably the belated fruit of long study of Bach's 'Forty-eight' and of more recent study of 'Cherubini's *traité*'.[1]

This new outlook is perceptible in the opening bars of the A flat Ballade, with its distribution of the principal melody over soprano, tenor, bass, and soprano successively. It leaves its marks even where one would least expect to find it: in the mazurkas—above all in Op. 50, No. 3, and in the beautiful canonic passage in the last twelve bars of Op. 63, No. 3. Indeed a tendency to canonic writing is one of the hall-marks of Chopin's third-period style (cf. the Ballade in F minor, bars 50–2, the first

[1] See p. xii.

movement of the Sonata in B minor, bars 23–8, etc.) Even when he does not employ contrapuntal devices, such as canon, Chopin betrays a strong inclination to think 'horizontally' in something like real parts instead of merely in broken up chords.

The tendency to evolve 'continuous' melodic lines through the decoration of a harmonic skeleton, already noticed in certain second-period works, now becomes more pronounced. The continuous melody of the Berceuse is something more than mere melodic ornamentation of the rocking chord basis; it has its own independent linear life, which is emphasized by the entry of the second part at bar 6. The G flat Impromptu, the latter part of the A flat Ballade, practically the whole of the F minor Ballade, a great deal of the Polonaise-Fantaisie, and the Fantaisie in F minor are rich in melodic lines 'continuous' in the Wagnerian rather than the Bachian sense, but very different from the discrete, stanzaic melodies of the earlier compositions. And, particularly in the two great F minor works, these melodic lines are often more virile, less sensuous than in the earlier Chopin. The Italian influence, as apparent as ever in the Berceuse and Barcarolle, here temporarily disappears, or is absorbed into something greater.

As we have just seen in the Berceuse, even where the Italian influence is as obvious as ever in the essential feeling, the 'continuous' tendency is often strongly pronounced. It is not only that the periods of the phrase-structure are still more skilfully disguised (compare the crucial ninth bar of the Nocturne, Op. 55, No. 2, with the first bar) or made more delightfully asymmetrical as in the Barcarolle and the *Più lento* section of the C sharp minor Valse, that the underlying harmony avoids marked cadences in a way that anticipates Wagner in method and often surpasses him in subtlety; the linear feeling itself is continuous (cf. in particular the Nocturnes, Op. 48, No. 2, Op. 55, No. 2, and Op. 62, No. 1, the Berceuse already mentioned, and even some of the mazurkas). The Nocturne, Op. 55, No. 2, also illustrates the principle of motive-generated melody in continuous *cantabile* form; compare bars 2, 3, and 14, and then again bar 14 with bars 19–24.

Coloratura ornamentation persists, too, though it appears less often. But now it is sometimes made thematic, as it was later made by Liszt in his B minor Sonata; the semiquaver passage in Ex. 39*b*, for instance, grows organically, by diminution, out of the first bar quoted.

This too is a symptom of the new quasi-symphonic tendency apparent in the music of the third period, a tendency manifest in the organic-thematic tightening up of links and ornaments, in the contrapuntal play with themes, and in a new sweep and power in the handling of large-scale form.

In his shorter works Chopin clung to the end to the ternary formula, though with ever greater subtlety of modification—beginning the repeat of the first section in the wrong key (Mazurka, Op. 59, No. 1), drastically varying it (Nocturne, Op. 48, No. 1), or both varying and shortening it

(Nocturnes, Op. 55, No. 1, and Op. 62, No. 1). In the Barcarolle the repeat of the first section includes a ten-bar quotation from the middle section.

The large-scale Mazurkas, Op. 50, No. 3, and Op. 56, Nos. 1 and 3, following the precedent of Op. 41, No. 1, described on page 49, have final development sections too important to be considered as mere codas, thus: ABA Development. (In the case of Op. 56, No. 1: ABAB′A Development, as the B section is repeated in a different key.) The F sharp minor Polonaise has not only a normal middle section but an interpolated mazurka[1]—AB Mazurka A—and the transition passage from mazurka back to polonaise is one of the finest examples of Chopin's last-period thematic workmanship. The return is made completely organic by the fusion of separate motives from the mazurka with the D–C sharp–F sharp–E sharp motive from which the whole-eight-bar introduction to the Polonaise has been built up:

Ex. 41

[1] In a letter of August, 1841, Chopin speaks of 'a new manuscript (a kind of polonaise, but it's more a fantasia)'—evidently the F sharp minor. The genre 'polonaise-fantaisie' was evidently in his mind some years before he wrote the work actually so called.

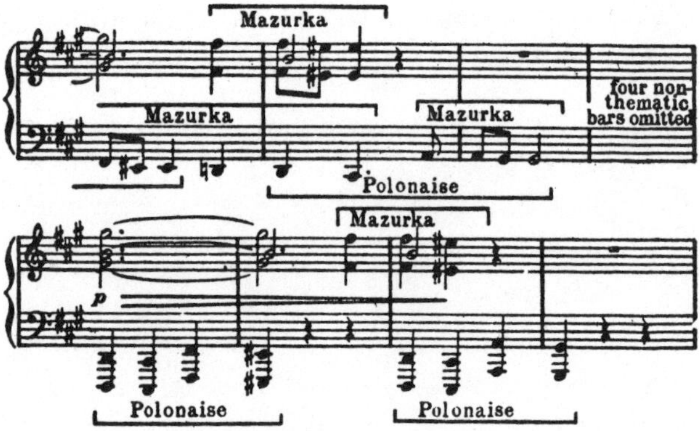

The introductory motive is then whipped up by diminution into its original form and the main subject of the Polonaise follows. The use of the theme at bar 9 of the Polonaise as a bass (at bar 17 and later) and the dying-away coda are hardly less masterly.

Still more remarkable is the Polonaise-Fantaisie, which ranks structurally with the E major Scherzo, the F minor Fantaisie, and the Third and Fourth Ballades as the crown of Chopin's work. Of these the Scherzo is the least interesting. It differs little in plan from its predecessors, Op. 31 and Op. 39, being laid out as follows:

Main section (two chief subjects in tonic and dominant)
Development
Main section (first subject only, in tonic but modulating to key of trio)
Trio (C sharp minor)

Main section (both subjects in original keys)
Development (as before) and Coda.

Similarly the only structural point of interest in the first movement of the B minor Sonata is merely a variation of a procedure in the B flat minor. This time the reprise does not omit the whole of the first subject, certainly, but it does skip the first sixteen bars of it, and drastically curtails the rest, so that, as Leichtentritt puts it, 'the broad singing melody of the second subject appears as the real goal of the long and complicated development' just as in the B flat minor; though with less force here, for it has not been excluded from the development as in the earlier Sonata.

But Chopin's conceptions of form and thematic development were, like those of his fellow romantics, Schumann and Berlioz, too radically different from Beethoven's and those of the earlier classic masters who had created the sonata, for him to be able to cast his ideas very successfully in this essentially classical form. While he had progressed very far from the pitiful would-be development of the early C minor Sonata and the non-thematic improvisatory pseudo-developments of the concertos, his developments—alike in the two mature piano sonatas and in the masterpieces we are about to discuss—are never comparable with the great classical models. They are essentially affairs of sequence, variation, and modulation, swept along by powerful winds of improvisatory inspiration and worked out with fine attention to detail. But here again Chopin must be judged not as an inferior successor of Beethoven but as the brilliant forerunner of Liszt and Wagner.

The great Ballade in A flat follows nearly the same structural plan as the G minor, but with three principal subjects instead of two, and with the development placed *after* the reprise of the second subject:

Exposition	A: in A flat B: F major and other keys C: A flat
Reprise:	B: A flat
Development:	Mostly on the B subject: various keys
Reprise (*continued*)	A: both in A flat and drastically C: curtailed

The development includes one striking symptom of Chopin's new attitude to thematic work. In the gigantic sequence through B major, C, D, and E flat with which it concludes, the theme of B is twice answered by the opening motive of A and finally fused with it into a single idea.

The form of the F minor Ballade is both simpler and more difficult to grasp. It is perhaps most easily explained as a masterly deformation of sonata form:

Introduction: 7 bars in the dominant key

First Subject: a 73-bar working of the 4-bar theme, treated as melody, bass, in canon, and so on

Second Subject: 19 bars, in B flat; even here the first subject appears once in the tenor
Development: 29 bars
Reprise of Introduction, ending in cadenza (A major)
Reprise of First Subject: beginning in D minor and modulating back to the tonic; treated first canonically and then with 'thematic coloratura' variation; shortened to 34 bars
Reprise of Second Subject: in D flat, and now extended and treated with more breadth
Improvisatory Transition: 20 bars, leading from D flat to C major
Coda: F minor, with entirely fresh treatment of the main theme

The great Fantaisie, Op. 49, is by no means fantasia-like in the sense of being loose or rambling in structure. After a slow, almost symphonic introductory march in F minor, a transitional idea in triplets (which recurs for transitional purposes throughout the work and again in the coda) leads to a 75-bar passage in which four main themes are stated in 'mediant' key-relationship: the first in F minor, the second in A flat major, the third in C minor, and the fourth in E flat major. The first two subjects are then repeated in C minor and G flat, and, with a new *lento* section in B major, constitute the middle section of the Fantaisie. All four subjects are then recapitulated so that their key-sequence (B flat minor, D flat, F minor, and A flat) now leads back to the relative major, in which key the piece ends with a coda that includes a reference to the *lento* theme of the middle section.

The Polonaise-Fantaisie is admittedly a harder nut to crack. There are at least five themes, three of which do not recur, and analysis of the form produces only this not very illuminating result:

Introduction: 23 bars, mainly on A: various keys
A: 42 bars; in A flat
B: 26 bars; A flat but modulating
A: 24 bars; A flat
C: 32 bars; B flat but modulating

Più lento { D: 33 bars; B major, etc
E: 34 bars in G sharp minor and B major—2 bars as in the Introduction—final 10-bar reference to E

Transition: 16 bars
A: 12 bars; A flat
D: 35 bars; A flat

Yet the sequence and contrast of musical events are perfectly satisfying and the whole piece rests solidly on its tonal pillars of A flat. Here, in fact, we have a free form resembling not so much the rather square-cut symphonic poems of Liszt as the symphonic poems of Richard Strauss written half a century later. The resemblance is increased by the final apotheosis of the themes I have labelled A and D, and by the thematic polyphony of such passages as bars 14–20 and 102–5.

Admittedly, the Polonaise-Fantaisie is, from the

point of view of form, a freakish exception in Chopin's work as a whole. But we must remember that it was one of his last compositions and that, had he lived, it might have been only the first of a new series of even bolder flights than he had taken yet. From the stylistic point of view it is his most advanced work, not only structurally but, as we have already seen, harmonically. Perhaps it is unkind to Liszt to remember that he said of it that 'although comprising thoughts which in beauty and grandeur equal—I would almost say surpass—anything Chopin has written, the work stands, on account of its pathological contents, outside the sphere of art'.

INDEX